Genetic Basic of Growth and Development

Genetic Basic of Growth and Development

Mahendra Kamalvanshi

RANDOM PUBLICATIONS
NEW DELHI (INDIA)

Genetic Basic of Growth and Development

ISBN 978-93-5111-593-9

Published in 2015 in India by

Reprint 2019

RANDOM PUBLICATIONS
4376-A/4B, Gali Murari Lal, Ansari Road
New Delhi-110 002
Phone : +9111-43580356, 011-23289044, 011-43142548
e-mail: sales@randompublications.com,
info@randompublications.com, randomexports@gmail.com

Type Setting by : Friends Media, Delhi-110089
Printed at : Mehra Printers, Delhi-110 092

Preface

Genetics is the study of genes, heredity, and variation in living organisms. It is generally considered a field of biology, but it intersects frequently with many of the life sciences and is strongly linked with the study of information systems. The father of genetics is Gregor Mendel, a scientist and Augustinian friar. Mendel studied 'trait inheritance,' patterns in the way traits were handed down from parents to offspring. He observed that organisms inherit traits by way of discrete "units of inheritance". This term, still used today, is a somewhat ambiguous definition of what is referred to as a gene. Trait inheritance and molecular inheritance mechanisms of genes are still a primary principle of genetics in the 21st century, but modern genetics has expanded beyond inheritance to studying the function and behavior of genes. Gene structure and function, variation, and distribution are studied within the context of the cell, the organism and within the context of a population. Genetics has given rise to a number of sub-fields including epigenetics and population genetics. Organisms studied within the broad field span the domain of life, including bacteria, plants, animals, and humans. Genetic processes work in combination with an organism's environment and experiences to influence development and behavior, often referred to as Nature versus nurture. The intra- or extra-cellular environment of a cell or organism may switch gene transcription on or off. A classic example is two seeds of genetically identical corn, one placed in a temperate climate and one in an arid climate. While the average height of the two corn stalks may be genetically determined to be equal, the one in the arid climate only grows to half the height of the one in the temperate climate, due to lack of water and nutrients in its environment.

I would like to thank my team for standing beside me throughout my career and writing this book. My special thanks go to "Random Publications" who have published the book.

– Mahendra Kamalvanshi

Contents

1

The Principles of Genetics

THE HEALING POWER OF DNA

Pharmacogenetics is advancing quickly since scientists have a lot of new information from the Human Genome Project and new computer tools that help them analyze the information. One disease for which progress has been rapid is cancer.

Consider the fact that cancer is often treated with a chemotherapy "cocktail," a combination of several different medicines. Each of the drugs in the mixture interacts with different proteins that control how well that particular drug works and how quickly it is metabolized in the body. What's more, each drug may have its own set of unpleasant—even potentially life-threatening—side effects. For these reasons, individually targeted, gene-based prescriptions for chemotherapy may offer a real benefit to people with cancer.

Currently, chemotherapy cures about 80 percent of the children who have been diagnosed with acute lymphoblastic leukemia, the most common childhood cancer. The remaining 20 percent are at risk of the cancer coming back. Mary Relling, a research clinical pharmacist at St. Jude Children's Research Hospital in Memphis, Tennessee, discovered that variations in two genes can predict which patients with acute lymphoblastic leukemia are likely to be cured by chemotherapy. Her research team also identified more than 100 genes expressed only in cancer cells that can be used to predict resistance to chemotherapy drugs. By taking patient and cancer cell genetic pro- files into account, Relling says, researchers can develop more effective treatments for the disease. Other pharmacogenetic scientists are studying the effects of gene variants on patients' responses to drugs used to treat AIDS, allergies, infections, asthma, heart disorders, and many other conditions.

For example, researchers recently identified two different genetic variants that play a central role in determining the body's response to Coumadin® (also known as warfarin), a widely prescribed medicine given to people who are at risk for blood clots or heart attacks. Although 2 million Americans take this blood-thinning drug every day, it is very difficult to administer, since its effects vary widely in different people who are taking the same dose. Giving the right dose is essential, because too much Coumadin can cause excessive bleeding, while too little can allow blood clots to form. Allan Rettie, a medicinal chemist at the University of Washington in Seattle, discovered that genetic variation among people influences the activity of a protein in the blood that is Coumadin's molecular target. He and other scientists are now trying to translate these findings into a genetic test that could help doctors predict what dose of Coumadin is appropriate based on each patient's DNA profile.

Genes in Human Behaviour

Did you know that, in addition to traits you can see like hair colour and physique, genes also contribute to how we behave? It may come as a surprise that many researchers are answering basic questions about the genetics of behaviour by studying insects.

For example, Gene Robinson, an entomologist at the University of Illinois at Urbana Champaign, works with honeybees. Robinson says that if you look at honeybees in their natural hive environment, you'll quickly see that they are very outgoing. In fact, according to Robinson, honeybees can't survive without the social structure of their community within the hive.

This characteristic makes them a perfect species in which to study the genetics of behaviour.What's particularly interesting about bees is that rather than being stuck in a particular job, they change jobs according to the hive's needs. Robinson has identified certain genes whose activity changes during a job shift, suggesting that the insects' environment helps to shape their gene expression.Researchers who are beginning to understand these connections are working in a brandnew field of investigation named by Robinson himself: sociogenomics.

What does all of this mean for humans, you wonder? It underscores the fact that, far from being set in stone, our genomes are influenced by both heredity and environment, finetuned and sculpted by our social life and the things we do every day.

Cause and Effect

What more do we need to know about how genes shape who we are and what we become?Context plays an enormous role in determining how organisms grow and develop, and what diseases they get. A unique combination of genetic and environmental factors, which interact in a way that is very hard to predict, determines what each person is like.Very few, if any, scientists would argue with this. Whether a gene is expressed, and even whether the mRNA transcript gets translated into a protein, depends on the environment. Few diseases—most of which are very rare—are caused completely by a mutated gene. In most cases, getting or avoiding a disease depends not just on genes but on things within your control, such as diet, exercise, and whether or not you smoke.It will be many years before scientists clearly understand the detailed meaning of our DNA language and how it interacts with the environment in which we live. Still, it's a great idea to find out as much as you can about your family's health history. Did any of your relatives have diabetes? Do people in your family tree have cancer or heart disease?

Keep in mind that diseases such as these are relatively common, so it's pretty likely that at least one relative will have one of them. But if heart disease, diabetes, or particular types of cancer "run in your family," especially if a lot of your relatives get the condition when they are fairly young, you may want to talk with your doctor about your own risk for developing the disease.If you do discover that you are at higher-than usual risk for a disease like breast cancer or heart disease, you may be able to prevent the disease, or delay its onset, by altering your diet, exercising more, or making other lifestyle changes. You may also be able to take advantage of screening tests like mammograms (breast X rays that detect signs of cancer), colonoscopies (imaging tests for colon cancer), or blood sugar tests for diabetes. Screening tests can catch diseases early, when treatment is most successful.

Many scientists focus on human genes, most of which have counterparts in the genomes of model organisms. However, in the case of infections caused by microorganisms, understanding how the genomes of bacteria, viruses, and parasites differ from ours is a very important area of health research.

Most of the medicines we take to treat infections by bacteria and viruses have come from scientists' search for molecular weak points in these tiny organisms. As mentioned earlier for example, some antibiotics kill bacteria by disarming their protein-making ribosomes.So why don't they kill human cells, too? The answer is that human and bacterial ribosomes are different. Genome sequencing is a powerful tool for identifying differences that might be promising targets for new drugs.

Comparing genetic sequences in organisms that are resistant and non-resistant to drugs can reveal new approaches to fighting resistance. Drug resistance is a worldwide problem for a number of diseases, including malaria.

Although researchers have developed several different types of medicines to treat this disease—caused by parasites carried by mosquitoes, not by a bacterium or a virus—malaria is rampant, especially in the developing world.This is partly because not all people have access to treatment, or to simple preventive measures like bed nets, which protect sleeping people from mosquito bites. But another problem is the malaria parasite itself, which has rapidly evolved ways to avoid the effects of antimalarial drugs.Scientists are trying to counter this process by studying microbial genetic information. In the case of malaria, geneticists like Dyann Wirth of the Harvard School of Public Health compare the genomes of drug-resistant parasites and those that can still be killed by antimalarial medicines. Wirth's research suggests that it should be possible to develop a simple, inexpensive genetic test that could be given to people with malaria, anywhere in the world. This test would identify drugs that are likely to be most effective and help decrease the rate at which parasites become resistant to the antimalarial medicines we already have.

These gangs, known as biofilms, are layers of slime that develop naturally when bacteria congregate on surfaces like stone, metal, and wood. Or on your teeth: yuck!Biofilms grow in all sorts of conditions. For example, one biofilm known as "desert varnish" thrives on rocks, canyon walls, or, sometimes, entire mountain ranges, leaving a reddish or other-colored stain. It is thought that petroglyphs left on boulders and cave walls by early desert dwellers were often formed by scraping through the coating of desert varnish formations with a hard object.

Sometimes, biofilms perform helpful functions. One of the best examples of the use of biofilms to solve an important problem is in the cleaning of wastewater.But biofilms can be quite harmful, contributing to a wide range of serious health problems including cholera, tuberculosis, cystic fibrosis, and food poisoning. They also underlie many conditions that are not life-threatening but are Non-etheless troublesome, like tooth decay and ear infections.

Bacteria form biofilms as a survival measure. By living in big groups rather than in isolation, the organisms are able to share nutrients and conserve energy. How do they do it?A biofilm is not just a loose clump of cells— it's a highly sophisticated structure. As in any community, the individuals in biofilms communicate with each other.Beyond that, many aspects of biofilms are poorly understood. Bacterial geneticist Bonnie Bassler of Princeton University in New Jersey is working to understand biofilms better, with the goal of being able to use this knowledge to break up bacterial "gang meetings."

Bassler's research subjects have a definite visual appeal. They glow in the dark, but only when they are part of a group. The bioluminescence, as the glow is called, arises from chemical reactions taking place within the biofilm. It provides a way for the bacteria to talk to each other, estimate the population size of their community, and distinguish themselves from other types of microorganisms.Through her studies, Bassler has identified a set of molecules that biofilm-forming microorganisms use to pass messages to each other. By devising genetically based methods to cut off the chatter, Bassler reasons, she may be able to cause bacterial communities to fall apart. This approach would provide a whole new way to treat health problems linked to harmful biofilms.

LIMITATIONS OF CURRENT TECHNOLOGY

Claims that the complete sequence of the human genome is available are somewhat inaccurate. For example, the Lawrence Berkeley Laboratory published a paper entitled "The complete sequence of human chromosome 5".

Careful inspection of this paper shows that despite the use of the word "complete", there are non-sequenced genomic regions, and that the authors can only estimate how much more work remains to be done. While some DNA regions, particularly those that are rich in simple sequence repeats such as telomeres, centromeres, and AT-rich intergenic regions are difficult or impossible to sequence with current technology, it is not clear how much of the unsequenced region on chromosome 5 resides in repeat-rich regions.

Researchers are beginning to use knowledge learned from genome sequencing research to Fig out how being healthy and being sick are different at the level of molecules. And doctors are starting to use genetic information to make treatment choices.For example, a diagnostic test can search for differences in the level of expression of a particular gene in breast cancer cells and predict whether a person will respond to a drug called Herceptin.The cancerous cells of some people who have breast cancer make an abundance of "HER2" proteins that are targeted by Herceptin. For those people, Herceptin is a miracle drug because it reduces the risk that their breast cancer will come back, and it also decreases their odds of dying from the disease.For cancer patients whose tumor genes do not express HER2, Herceptin won't do a thing, though, so it shouldn't be prescribed. Research is proceeding quickly to develop other genetic tests that may help diagnose and treat a wide range of health problems beyond cancer.

GENETIC TECHNOLOGIES

A variety of techniques exist for manipulating DNA in the laboratory. Restriction enzymes are a commonly used enzyme that cuts DNA at specific sequences, producing predictable fragments of DNA. The use of ligation enzymes allows these fragments to be stitched back together, and by ligating fragments of DNA together from different sources, researchers can create recombinant DNA. Often associated with genetically modified organisms, recombinant DNA is commonly used in the context of plasmids — short circular DNA fragments with a few genes on them. By inserting plasmids into bacteria and growing those bacteria on plates of agar (to isolate clones of bacteria cells), researchers can clonally amplify the inserted fragment of DNA (a process known as molecular cloning). (Cloning can also refer to the creation of clonal organisms, through various techniques.)

DNA can also be amplified using a procedure called the polymerase chain reaction (PCR). By using specific short sequences of DNA, PCR can exponentially amplify a targeted region of DNA. Because it can amplify from extremely small amounts of DNA, PCR is often used to detect the presence of specific DNA sequences.

MODEL ORGANISMS AND GENETICS

Although geneticists originally studied inheritance in a wide range of organisms, researchers began to specialize in studying the genetics of a particular subset of organisms. The fact that significant research already existed for a given organism would encourage new researchers to choose it for further study, and so eventually a few "model organisms" became the basis for most genetics research.

Common research topics in model organism genetics include the study of gene regulation and the involvement of genes in development and cancer.Organisms were chosen, in part, for convenience — short generation times and facile genetic manipulation made some organisms popular genetics research tools. Widely used model organisms include the gut bacterium Escherichia coli, the plant Arabidopsis thaliana, baker's yeast (Saccharomyces cerevisiae), the nematode Caenorhabditis elegans, the common fruit fly (Drosophila melanogaster), and the common house mouse (Mus musculus).

MENDEL'S LAWS

Law of Segregation

Mendel studied the inheritance of seed shape first.A cross involving only one trait is referred to as a monohybrid cross. Mendel crossed pure-breeding (also referred to as true-breeding) smooth-seeded plants with a variety that had always produced wrinkled seeds (60 fertilizations on 15 plants). All resulting seeds were smooth.

The following year, Mendel planted these seeds and allowed them to self-fertilize. He recovered 7324 seeds: 5474 smooth and 1850 wrinkled. To help with record keeping, generations were labeled and numbered. The parental generation is denoted as the P1 generation. The offspring of the P1 generation are the F1 generation (first filial). The selffertilizing F1 generation produced the F2 generation (second filial).

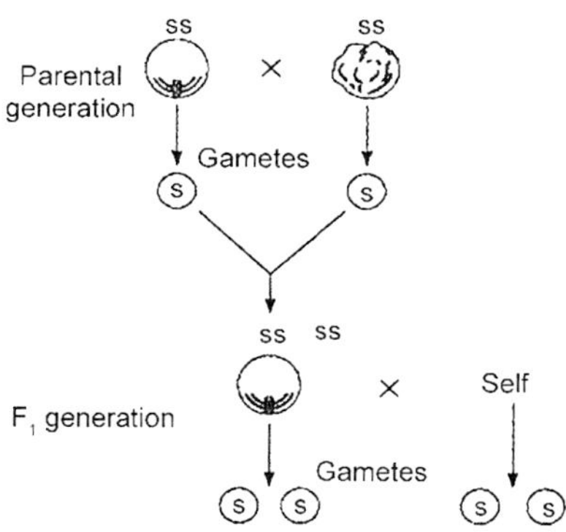

Fig. Law of Segregation.

The Law of Segregation, also known as Mendel's First Law, essentially has four parts.

1. Alternative versions of genes account for variations in inherited characteristics. This is the concept of alleles. Alleles are different versions of genes that impart the same characteristic. For example, each human has a gene that controls eye colour, but there are variations among these genes in accordance with the specific colour the gene "codes" for.
2. For each characteristic, an organism inherits two alleles, one from each parent. This means that when somatic cells are produced from two gametes, one allele comes from the mother and one from the father. These alleles may be the same (true-breeding organisms/ homozygous *e.g.* ww and rr or different
3. If the two alleles differ, then one, the allele that encodes the dominant trait, is fully expressed in the organism's appearance; the other, the allele encoding the recessive trait, has no noticeable effect on the organism's appearance. In other words, only the dominant trait is seen in the phenotype of the organism. This allows recessive traits to be passed on to offspring even if they are not expressed. Not all traits have a dominant-recessive relationship, however. The "Japanese wonder flower" Mirabilis jalapa illustrates incomplete dominance. There is also codominance *e.g.* Human blood types where A and B are codominant and O is recessive.
4. The two alleles for each characteristic segregate during gamete production. This means that each gamete will contain only one allele for each gene. This allows the maternal and paternal alleles to be combined in the offspring, ensuring variation.

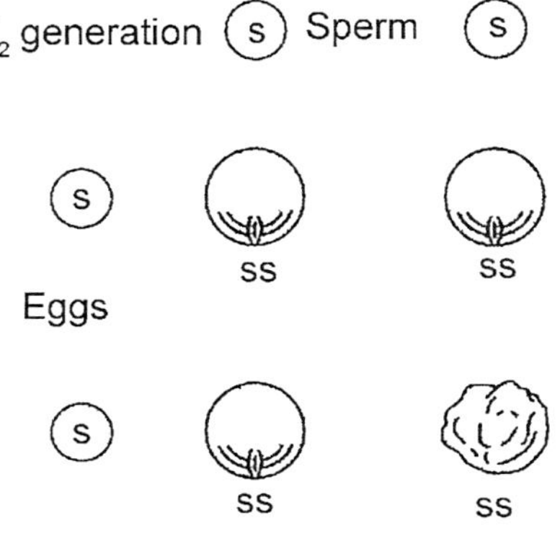

Fig. Punnett square

Punnett square explaining the behaviour of the S and s alleles.

P1: smooth X wrinkled

F1: all smooth

F2: 5474 smooth and 1850 wrinkled

Meiosis, a process unknown in Mendel's day, explains how the traits are inherited.

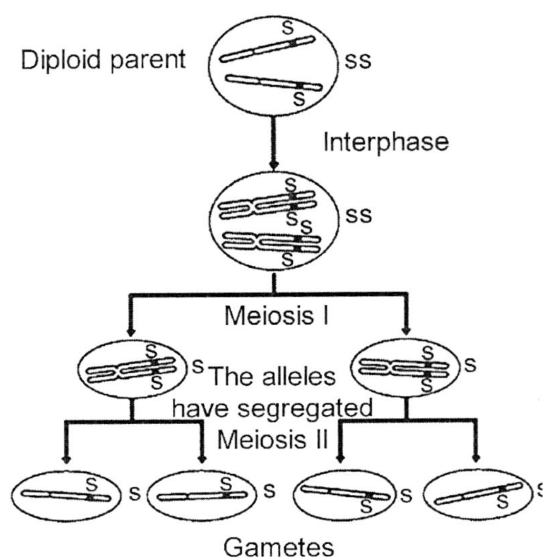

Fig. The inheritance of the S and s alleles

N.B It is often misconstrued that the gene itself is dominant, recessive, codominant, or incompletely dominant. It is, however the trait, or gene product that the allele encodes that is dominant, etc.

Law of Independent Assortment

The Law of Independent Assortment, also known as Mendel's Second Law, states that the emergence of one trait will not affect the emergence of another. While his experiments with mixing one trait always resulted in a 3:1 ratio between dominant and recessive phenotypes, his experiments with mixing two traits (dihybrid cross) showed 9:3:3:1 ratios. Mendel concluded that different traits are inherited independently of each other, so that there is no relation, for example, between a cat's colour and tail length. This is actually only true for genes that are not linked to each other.

1. Parental generation 2. F_1 generation.
3. F_2 generation. Dominant (red) and recessive (white) phenotype look alike in the F_1 (first) generation and show a 3:1 ratio in the F_2 (second) generation

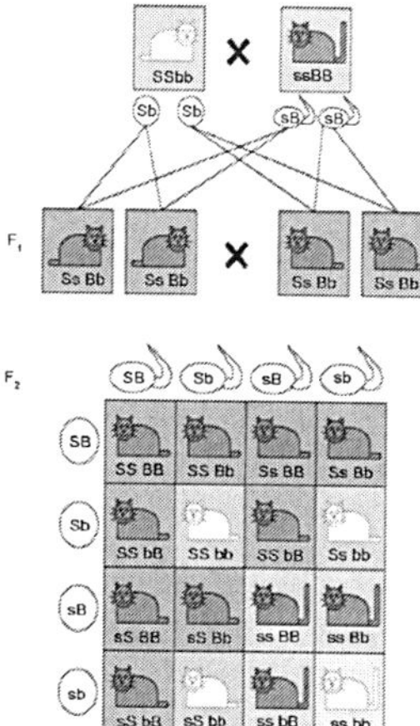

Fig: The genotypes of two independent traits show a 9:3:3:1 ratio in the F_2 generation. In this example, coat colour is indicated by B (brown, dominant) or b (white) while tail length is indicated by S (short, dominant) or s (long). When parents are homozygous for each trait ('SSbb and ssBB), their children in the F_1 generation are heterzygous at both loci and only show the dominant phenotypes. If the children mate with each other, in the F_2 generation all combination of coat colour and tail length occur: 9 are brown/short (purple boxes), 3 are white/short (pink boxes), 3 are brown/long (blue boxes) and 1 is white/long (green box).

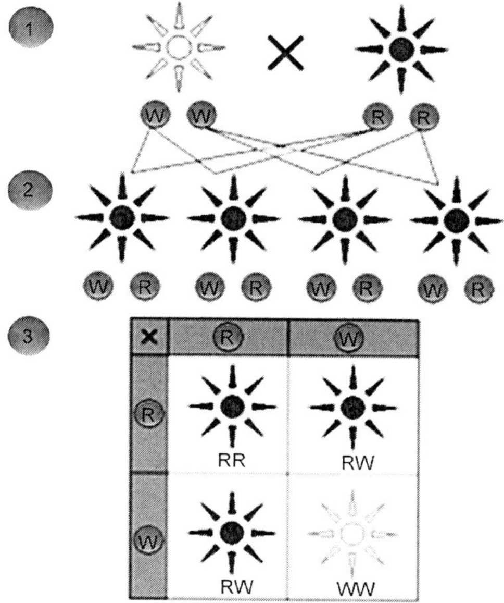

Fig: The colour alleles of Mirabilis jalapa are not dominant or recessive.

1. Parental generation.
2. F_1 generation.

3. F_2 generation. The "red" and "white" allele together make a "pink" phenotype, resulting in a 1:2:1 ratio of red:pink:white in the F_2 generation.

The reason for these laws is found in the nature of the cell nucleus. It is made up of several chromosomes carrying the genetic traits. In a normal cell, each of these chromosomes has two parts, the chromatids. A reproductive cell, which is created in a process called meiosis, usually contains only one of those chromatids of each chromosome. By merging two of these cells (usually one male and one female), the full set is restored and the genes are mixed. The resulting cell becomes a new embryo.

The fact that this new life has half the genes of each parent is one reason for the Mendelian laws. The second most important reason is the varying dominance of different genes, causing some traits to appear unevenly instead of averaging out (whereby dominant doesn't mean more likely to reproduce - recessive genes can become the most common, too).There are several advantages of this method (sexual reproduction) over reproduction without genetic exchange:

1. Instead of nearly identical copies of an organism, a broad range of offspring develops, allowing more different abilities and evolutionary strategies.
2. There are usually some errors in every cell nucleus. Copying the genes usually adds more of them. By distributing them randomly over different chromosomes and mixing the genes, such errors will be distributed unevenly over the different children. Some of them will therefore have only very few such problems. This helps reduce problems with copying errors somewhat.
3. Genes can spread faster from one part of a population to another. This is for instance useful if there's a temporary isolation of two groups. New genes developing in each of the populations don't get reduced to half when one side replaces the other, they mix and form a population with the advantages of both sides.
4. Sometimes, a mutation (e. g. Sickle Cell Anemia) can have positive side effects (in this case Malaria resistance). The mechanism behind the Mendelian laws can make it possible for some offspring to carry the advantages without the disadvantages until further mutations solve the problems.

MENDELIAN TRAIT

A Mendelian trait is one that is controlled by a single locus and shows a simple Mendelian inheritance pattern. In such cases, a mutation in a single gene can cause a disease that is inherited according to Mendel's laws. Examples include sickle-cell anemia, Tay-Sachs disease, cystic fibrosis and xeroderma pigmentosa. A disease controlled by a single gene contrasts with a multi-factorial disease, like arthritis, which is affected by several loci (and the environment) as well as those diseases inherited in a non-Mendelian fashion. The Mendelian Inheritance in Man database is a catalog of, among other things, genes in which Mendelian mutants causes disease.Mendel studied seven traits which appeared in two discrete forms, rather than continuous characters which are often difficult to distinguish. When “truebreeding” tall plants were crossed with “true-breeding” short plants, all of the offspring were tall plants. The parents in the cross were the P1 generation, and the offspring represented the F1 generation.

The trait referred to as tall was considered dominant, while short was recessive. Dominant traits were defined by Mendel as those which appeared in the F1 generation in crosses between true-breeding strains. Recessives were those which “skipped” a generation, being expressed only when the dominant trait is absent. Mendel’s plants exhibited complete dominance, in which the phenotypic expression of alleles was either dominant or recessive, not “in between”. When members of the F1 generation were crossed, Mendel recovered mostly tall offspring, with some short ones also occurring.

Upon statistically analyzing the F2 generation, Mendel determined the ratio of tall to short plants was approximately 3:1. Short plants have skipped the F1 generation, and show up in the F2 and succeeding generations. Mendel concluded that the traits under study were governed by discrete (separable) factors. The factors were inherited in pairs, with each generation having a pair of trait factors. We now refer to these trait factors as alleles. Having traits inherited in pairs allows for the observed phenomena of traits “skipping” generations.

Summary of Mendel’s Results:

1. The F1 offspring showed only one of the two parental traits, and always the same trait.
2. Results were always the same regardless of which parent donated the pollen (was male).
3. The trait not shown in the F1 reappeared in the F2 in about 25per cent of the offspring.
4. Traits remained unchanged when passed to offspring: they did not blend in any offspring but behaved as separate units.
5. Reciprocal crosses showed each parent made an equal contribution to the offspring.

Mendel’s Conclusions:

1. Evidence indicated factors could be hidden or unexpressed, these are the recessive traits.
2. The term phenotype refers to the outward appearance of a trait, while the term genotype is used for the genetic makeup of an organism.
3. Male and female contributed equally to the offsprings’ genetic makeup: therefore the number of traits was probably two (the simplest solution).

4. Upper case letters are traditionally used to denote dominant traits, lower case letters for recessives.

Mendel reasoned that factors must segregate from each other during gamete formation (remember, meiosis was not yet known!) to retain the number of traits at 2. The Principle of Segregation proposes the separation of paired factors during gamete formation, with each gamete receiving one or the other factor, usually not both. Organisms carry two alleles for every trait. These traits separate during the formation of gametes.

Dihybrid Crosses

When Mendel considered two traits per cross (dihybrid, as opposed to singletrait-crosses, monohybrid), The resulting (F2) generation did not have 3:1 dominant:recessive phenotype ratios. The two traits, if considered to inherit independently, fit into the principle of segregation. Instead of 4 possible genotypes from a monohybrid cross, dihybrid crosses have as many as 16 possible genotypes.Mendel realized the need to conduct his experiments on more complex situations. He performed experiments tracking two seed traits: shape and colour. A cross concerning two traits is known as a dihybrid cross.

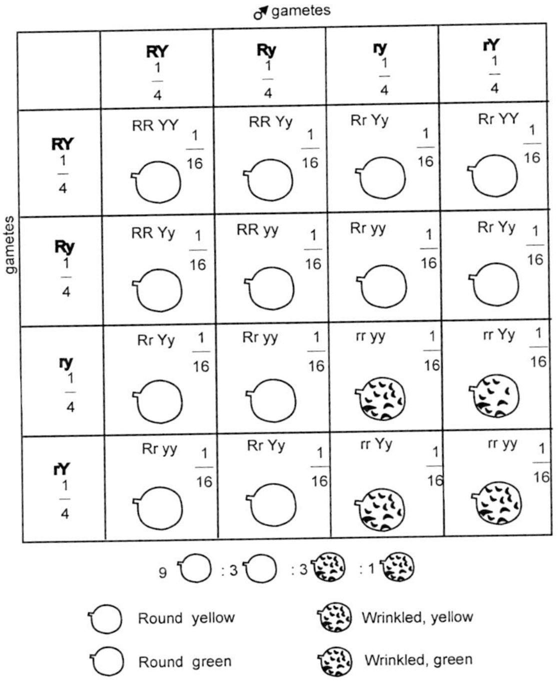

Fig. Dihybrid Cross

Crosses With Two Traits

- Smooth seeds (S) are dominant over wrinkled (s) seeds.

- Yellow seed colour (Y) is dominant over green (g).

Methods, Results, and Conclusions

Mendel started with true-breeding plants that had smooth, yellow seeds and crossed them with true-breeding plants having green, wrinkled seeds. All seeds in the F1 had smooth yellow seeds. The F2 plants self-fertilized, and produced four phenotypes:

315 smooth yellow
108 smooth green
101 wrinkled yellow
32 wrinkled green

Mendel analyzed each trait for separate inheritance as if the other trait were not present.The 3:1 ratio was seen separately and was in accordance with the Principle of Segregation. The segregation of S and s alleles must have happened independently of the segregation of Y and y alleles. The chance of any gamete having a Y is 1/2; the chance of any one gamete having a S is 1/ 2.The chance of a gamete having both Y and S is the product of their individual chances (or 1/2 X 1/2 = 1/4). The chance of two gametes forming any given genotype is 1/4 X 1/4 (remember, the product of their individual chances). Thus, the Punnett Square has 16 boxes. Since there are more possible combinations to produce a smooth yellow phenotype (SSYY, SsYy, SsYY, and SSYy), that phenotype is more common in the F2.

From the results of the second experiment, Mendel formulated the Principle of Independent Assortment — that when gametes are formed, alleles assort independently. If traits assort independent of each other during gamete formation, the results of the dihybrid cross can make sense. Since Mendel's time, scientists have discovered chromosomes and DNA. We now interpret the Principle of Independent Assortment as alleles of genes on different chromosomes are inherited independently during the formation of gametes. This was not known to Mendel.

Punnett squares deal only with probability of a genotype showing up in the next generation. Usually if enough offspring are produced, Mendelian ratios will also be produced.

Step 1 - definition of alleles and determination of dominance.

Step 2 - determination of alleles present in all different types of gametes.

Step 3 - construction of the square.

Step 4 - recombination of alleles into each small square.

Step 5 - Determination of Genotype and Phenotype ratios in the next generation.

Step 6 - Labeling of generations, for example P1, F1, etc.

While answering genetics problems, there are certain forms and protocols that will make unintelligible problems easier to do. The term "true-breeding strain" is a code word for homozygous. Dominant alleles are those that show up in the next generation in crosses between two different "true-breeding strains". The key to any genetics problem is the recessive phenotype (more properly the phenotype that represents the recessive genotype). It is that organism whose genotype can be determined by examination of the phenotype. Usually homozygous dominant and heterozygous individuals have identical phenotypes (although their genotypes are different). This becomes even more important in dihybrid crosses.

Assortment and interactions of multiple genes

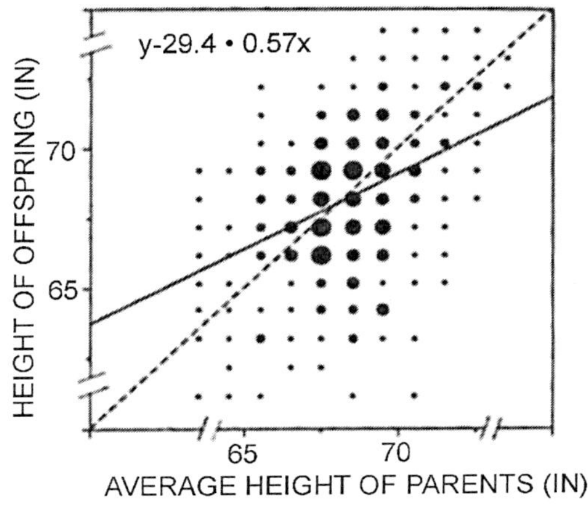

Fig. Law of independent assortment

Human height is a complex genetic trait. Francis Galton's data from 1889 shows the relationship between offsping height as a function of mean parent height. While correlated, the remaining variation in offspring heights indicates environment is also an important factor in this trait.

Organisms have thousands of genes, and in diploid organism assortment of these genes are generally independent of each other. This means that the inheritance of an allele for yellow or green pea colour is unrelated to the inheritance of alleles for white or purple flowers. This phenomenon, known as "Mendel's second law" or the "Law of independent assortment", means that the alleles of different genes get shuffled between parents to form children with many different combinations.

Often different genes can interact in a way that influences the same trait. In the blue-eyed Mary, for example, there exists a gene with alleles that determine the colour of flowers: blue or magenta. Another gene, however, controls whether the flowers have colour at all: colour or white. When a plant has two copies of this white allele, its flowers are white regardless of whether the first gene has blue or magenta alleles. This interaction between genes is called "epistasis", with the second gene epistatic to the first Many traits are not discrete features (eg. purple or white flowers) but are instead continuous features (eg. human height and skin colour). These "complex traits" are the product of interactions of many genes. The influence of these genes is mediated, to varying degrees, by the enviroment an organism has experienced. The degree to which an organism's genes contribute to a complex trait is called "heritability". Measurement of the heritability of a trait is relative, though — in a more variable environment, the environment has a bigger influence on the total variation of the trait. For example, human height is a complex trait with a heritability of 89per cent in the United States. In Nigeria, however, where people experience a more variable access to good nutrition and health care, height has a heritability of only 62per cent.

DNA and the genetic code

The molecular basis for genes is deoxyribonucleic acid (DNA). DNA is composed of a chain of nucleotides, of which there are four types: adenine (A), cytosine (C), guanine (G), and thymine (T). Genetic information exists in the sequence of these nucleotides, and genes exist as stretches of sequence along the DNA chain. Viruses are a small exception — the similar molecule RNA instead of DNA is often the genetic material of a virus. In all non-virus organisms, which are composed of cells, each cell contains a full copy of that organism's DNA, called its genome.

In the cell, DNA exists as a double-stranded molecule, coiled into the shape of a doublehelix. Each nucleotide in DNA preferentially pairs with its partner nucleotide on the opposite strand: A pairs with T, and C pairs with G. Thus, in its two-stranded form, each strand effectively contains all necessary information, redundant with its partner strand. This structure of DNA is the physical basis for inheritance: DNA replication duplicates the genetic information by splitting the strands and using each strand as a template for a partner strand.

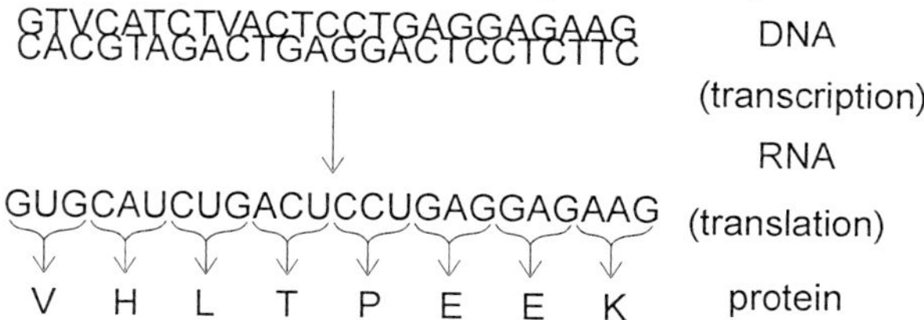

Fig. The genetic code: DNA, through a messenger RNA intermediate, codes for protein with a triplet code.

Genes express their functional effect through the production of proteins, which are complex molecules responsible for most functions in the cell. Proteins are chains of amino acids, and the DNA sequence of a gene (through an RNA intermediate) is used to produce a specific protein sequence. Each group of three nucleotides in the sequence, called a codon, corresponds to one of the twenty possibly amino acids in protein — this correspondence is called the genetic code. The specific sequence of amino acids results in a unique three-dimensional structure for that protein, thereby determining its behaviour and function.

MENDELIAN INHERITENCE

Gregor Johann Mendel was a monk in the Augustinian Monastery in the Brunn, Czech Republic. In 1854 he began the experiments which started modern genetics. His work with garden peas, Pisum sativum, was vital to our understanding of inheritance. He is known as the Father of Genetics.

MENDEL'S EXPERIMENT

Mendel reasoned an organism for genetic experiments should have:

1. A number of different traits that can be studied
2. Plant should be self-fertilizing and have a flower structure that limits accidental contact
3. Offspring of self-fertilized plants should be fully fertile.

Mendel's experimental organism was a common garden pea (Pisum sativum), which has a flower that lends itself to self-pollination. The male parts of the flower are termed the anthers. They produce pollen, which contains the male gametes (sperm). The female parts of the flower are the stigma, style, and ovary. The egg (female gamete) is produced in the ovary. The process of pollination (the transfer of pollen from anther to stigma) occurs prior to the opening of the pea flower. The pollen grain grows a pollen tube which allows the sperm to travel through the stigma and style, eventually reaching the ovary. The ripened ovary wall becomes the fruit (in this case the pea pod). Most flowers allow cross-pollination, which can be difficult to deal with in genetic studies if the male parent plant is not known. Since pea plants are selfpollinators, the genetics of the parent can be more easily understood. Peas are also self-compatible, allowing self-fertilized embryos to develop as readily as out-fertilized embryos. Mendel tested all 34 varieties of peas available to him through seed dealers. The garden peas were planted and studied for eight years. Each character studied had two distinct forms, such as tall or short plant height, or smooth or wrinkled seeds. Mendel's experiments used some 28,000 pea plants.

Mendel's first step was breeding pure breeding strains of peas. The traits he studied included:

- Pea programmes
- Height
- And whether the Peas were wrinkled or smooth.

Mendel crossed the pure breeding Parental Generation (designated P). He found that the first generation (F1) was exclusively phenotypically one of the parental types. Mendel then crossed his F1 generation with itself. He found that the F2 generation showed a surprising trait, three quarters were like the F1 generation, while the remaining quarter were like the other Parents.From this Mendel realised that there were two versions of each loci, one of which expressed dominance over the other. He called this Biparticulate Inheritance. If a gene was following this 3:1 pattern it was said to be segregating normally. By looking at multiple genes, Mendel showed that they were not linked to each other and that each loci he studied had no influence over the others. He called this Independent assortment.By studying cases where the Mendelian laws we can also learn a lot. For instance, if a gene isn't segragating normally it may be sex linked. If two genes aren't Assorting Independantly they're probably on the same chromosome.Mendel's laws are the first step to understanding Genetics, they lay down the basic concept of inheritance.

Mendel's contribution was unique because of his methodical approach to a definite problem, use of clear-cut variables and application of mathematics (statistics) to the problem. Gregor Using pea plants and statistical methods, Mendel was able to demonstrate that traits were passed from each parent to their offspring through the inheritance of genes.

FEATURES OF INHERITANCE

Mendelian inheritance (or Mendelian genetics or Mendelism) is a set of primary tenets relating to the transmission of hereditary characteristics from parent organisms to their children; it underlies much of genetics. They were initially derived from the work of Gregor Mendel published in 1865 and 1866 which was "re-discovered" in 1900, and were initially very controversial. When they were integrated with the chromosome theory of inheritance by Thomas Hunt Morgan in 1915, they became the core of classical genetics.

Mendel's work showed:

1. Each parent contributes one factor of each trait shown in offspring.
2. The two members of each pair of factors segregate from each other during gamete formation.
3. The blending theory of inheritance was discounted.
4. Males and females contribute equally to the traits in their offspring.
5. Acquired traits are not inherited.

EXCEPTIONS TO SIMPLE INHERITANCE

Since Mendel's time, our knowledge of the mechanisms of genetic inheritance has grown immensely. For instance, it is now understood that inheriting one allele can, at times, increase the chance of inheriting another or can affect how and when a trait is expressed in an individual's phenotype. Likewise, there are degrees of dominance and recessiveness with some traits. The simple rules of Mendelian inheritance do not apply in these and other exceptions.

Polygenic Traits

Some traits are determined by the combined effect of more than one pair of genes. These are referred to as polygenic or continuous traits. An example of this is human stature. The combined size of all of the body parts from head to foot determines the height of an individual. There is an additive effect. The sizes of all of these body parts are, in turn, determined by numerous genes. Human skin, hair, and eye colour are also polygenic traits because they are influenced by more than one allele at different loci. The result is the perception of continuous gradation in the expression of these traits.

Intermediate Expression

Apparent blending can occur in the phenotype when there is incomplete dominance resulting in an intermediate expression of a trait in heterozygous individuals.

For instance, in primroses, snapdragons, and four-o'clocks, red or white flowers are homozygous while pink ones are heterozygous. The pink flowers result because the single "red" allele is unable to code for the production of enough red pigment to make the petals dark red.

homozygous dominant (AA)

heteroxygous (Aa)

Homoxygous recessive (Aa)

Another example of an intermediate expression may be the pitch of human male voices. The lowest and highest pitches apparently are found in men who are homozygous for this trait (AA and aa), while the intermediate range baritones are heterozygous (Aa). The child-killer disease known as Tay-Sachs is also characterized by incomplete dominance. Heterozygous individuals are genetically programmed to produce only 40-60per cent of the normal amount of an enzyme that prevents the disease.

Fortunately for Mendel, the pea plant traits that he studied were controlled by genes that do not exhibit an intermediate expression in the phenotype. Otherwise, he probably would not have discovered the basic rules of genetic inheritance.

Codominance

For some traits, two alleles can be codominant. That is to say, both are expressed in heterozygous individuals. An example of this is people who have an AB blood type for the ABO blood system. When they are tested, these individuals actually have the characteristics of both type A and type B blood. Their phenotype is not intermediate between the two.

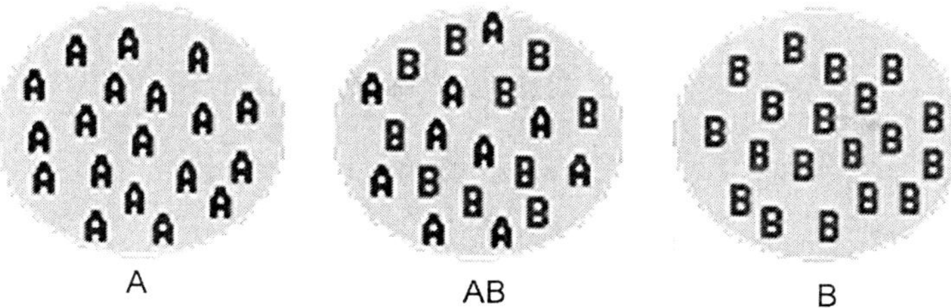

Multiple-allele Series

The ABO blood type system is also an example of a trait that is controlled by more than just a single pair of alleles. In other words, it is due to a multiple-allele series. In this case, there are three alleles (A, B, and O), but each individual only inherits two of them (one from each parent).

Some traits are controlled by far more alleles. For instance, the human HLA system, which is responsible for identifying and rejecting foreign tissue in our bodies, can have at least 30,000,000 different genotypes. It is the HLA system which causes the rejection of organ transplants. The more we learn about human genetics the more it becomes clear that multiple-allele series are very common. In fact, it now appears that they are more common than simple two allele ones.

DOMINANT AND RECESSIVE CHARACTERS

Substitution has been made in the case of the two kinds of daughters, with the same effect. Wherever an X-chromosome from the original father (white) is substituted for one from the original mother (black), the animal has bar-eyes. Where- ever an X-chromosome from the mother (black) is substituted for one from the father (white), the individual has normal eyes instead of bar-eyes.

It is certain therefore that bar-eye is an abnormality resulting from a defect in certain X-chromosomes. It is clear that X-chromosomes can become defective, and that this change in the X-chromosome causes the individual to develop in an abnormal way, so that it possesses a bodily abnormalityÿÿin this case, an abnormality of the eye. This abnormality appears, as above seen, in any individual that has the defective X-chromosome, whether it also has a normal chromosome or not.

INDIVIDUALIZED PRESCRIPTIONS

One way variations in our genes make a difference in our health is by affecting how our bodies react to medicines. The unsettling truth is that medicines work as expected in fewer than half of the people who take them.

While environmental and lifestyle factors can explain some of this, a good part of the individual variability in response to medicines can be attributed to variants in the genes that make cytochrome P450 proteins. These proteins process many of the drugs we take.

Because each person's set of genes is a little different, the proteins that the genes encode are also slightly different. These changes can affect how the cytochrome P450 proteins (and many other types of proteins) work on drugs.

Doctors first realized this in the 1950s, when some patients had bad—sometimes fatal—reactions to an anesthetic medicine used in surgery. Experiments revealed that those who reacted poorly had a genetic variation in the enzyme that breaks down and disposes of the anesthetic after it's been in the body for a while.

People whose genes encode the variant enzyme had no trouble at all until they needed surgery that required general anesthesia. In the operating room, a normal human genetic variation suddenly led to a medical crisis!

Fortunately, this type of serious reaction to an anesthetic is very rare. But many reactions to medicines aren't so unusual. Researchers know that genetic variations can cause some common medicines to have dangerous side effects. For example, some people who take the colon cancer drug Camptosar® (also known as irinotecan) can develop diarrhea and a life-threatening infection if they have a variant form of the gene for the protein that metabolizes Camptosar.

Genetic variations can also cause drugs to have little effect at all. For example, in some people, pain medicines containing codeine, like Tylenol® with Codeine Elixir, offer no relief because their bodies break it down in an unusual way.

The use of genetic information to predict how people will respond to medicines is called pharmacogenetics. The ultimate goal of this field of study is to customise treatments based on an individual's genes.

With this kind of approach, every patient won't be treated the same, because doctors will have the molecular tools to know ahead of time which drug, and how much of it, to prescribe—or whether to prescribe it at all.

DNA SEQUENCING AND GENOMICS

One of the most fundamental technologies developed to study genetics, DNA sequencing allows researchers to determine the sequence of nucleotides in DNA fragments. Developed in 1978 by Frederick Sanger and coworkers, chain-termination sequencing is now routinely used to sequence DNA fragments. With this technology, researchers have been able to study the molecular sequences associated with many human diseases. As sequencing has become less expensive and with the aid of computational tools, researchers have sequenced the genomes of many organisms by stitching together the sequences of many different fragments (a process called "genome assembly"). The Human genome project was completed in 2005, sequencing the genome to 92per cent coverage.The large amount of sequences available has created the field of "genomics", research which uses computational tools to search for and analyze patterns in the full genomes of organisms. Genomics can also be considered a subfield of bioinformatics, which uses computational approaches to analyze of large sets of biological data.

DNA sequencing

The term DNA sequencing encompasses biochemical methods for determining the order of the nucleotide bases, adenine, guanine, cytosine, and thymine, in a DNA oligonucleotide. The sequence of DNA constitutes the heritable genetic information in nuclei, plasmids, mitochondria, and chloroplasts that forms the basis for the developmental programmes of all living organisms.

Determining the DNA sequence is therefore useful in basic research studying fundamental biological processes, as well as in applied fields such as diagnostic or forensic research. The advent of DNA sequencing has significantly accelerated biological research and discovery. The rapid speed of sequencing attainable with modern DNA sequencing technology has been instrumental in the large-scale sequencing of the human genome, in the Human Genome Project. Related projects, often by scientific collaboration across continents, have generated the complete DNA sequences of many animal, plant, and microbial genomes.

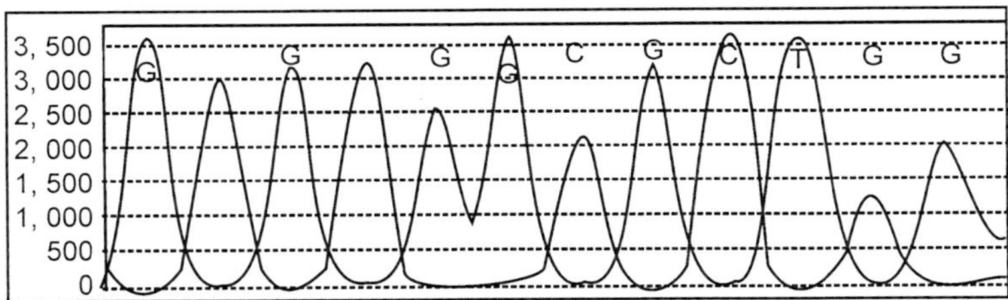

Fig.DNA Sequence Trace

Early methods

For thirty years, a large proportion of DNA sequencing has been carried out with the chain-termination method, developed by Frederick Sanger and coworkers in 1975. Prior to the development of rapid DNA sequencing methods in the early 1970s by Sanger in England and Gilbert et al. at Harvard,, a number of laborious methods were used. For instance, in 1973 Gilbert and Maxam reported the sequence of 24 basepairs using a method known as wandering-spot analysis.

It is noteworthy that RNA sequencing, which for technical reasons is easier to perform than DNA sequencing, could be considered one of the earliest forms of nucleotide sequencing. The major landmark of RNA sequencing, dating from the pre-recombinant DNA era, is the sequence of the phage MS2 genome, identified and published by Walter Fiers and coworkers.

Maxam-Gilbert sequencing

In 1976-1977, Allan Maxam and Walter Gilbert developed a DNA sequencing method based on chemical modification of DNA and subsequent cleavage at specific bases. Although Maxam and Gilbert published their chemical sequencing method two years after the ground-breaking paper of Sanger and Coulson on plus-minus sequencing, Maxam-Gilbert sequencing rapidly became more popular, since purified DNA could be used directly, while the initial Sanger method required that each read start be cloned for production of single-stranded DNA. However, with the development and improvement of the chain-termination method, Maxam-Gilbert sequencing has fallen out of favour due to its technical complexity, extensive use of hazardous chemicals, and difficulties with scale-up. In addition, unlike the chain-termination method, chemicals used in the Maxam Gilbert method cannot easily be customized for use in a standard molecular biology kit.

In brief, the method requires radioactive labelling at one end and purification of the DNA fragment to be sequenced. Chemical treatment generates breaks at a small proportion of one or two of the four nucleotide bases in each of four reactions (G, A+G, C, C+T). Thus a series of labelled fragments is generated, from the radiolabelled end to the first 'cut' site in each molecule. The fragments are then size-separated by gel electrophoresis, with the four reactions arranged side by side. To visualize the fragments generated in each reaction, the gel is exposed to X-ray film for autoradiography, yielding an image of a series of dark 'bands' corresponding to the radiolabelled DNA fragments, from which the sequence may be inferred.

Also sometimes known as 'chemical sequencing', this method originated in the study of DNA-protein interactions (footprinting), nucleic acid structure and epigenetic modifications to DNA, and within these it still has important applications.

Chain-termination Methods

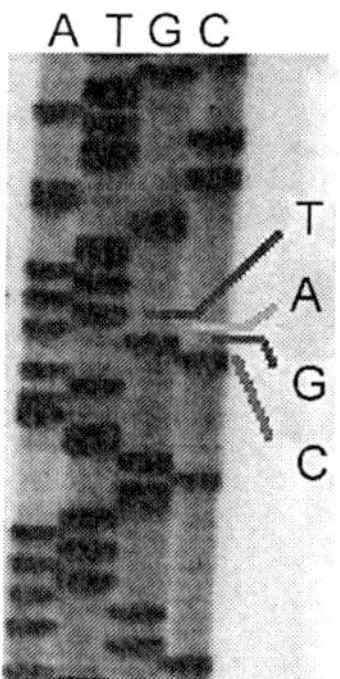

Fig. Part of a radioactively labelled sequencing gel

While the chemical sequencing method of Maxam and Gilbert, and the plus-minus method of Sanger and Coulson were orders of magnitude faster than previous methods, the chain-terminator method developed by Sanger was even more efficient, and rapidly became the method of choice. The Maxam-Gilbert technique requires the use of highly toxic chemicals, and large amounts of radiolabeled DNA, whereas the chain-terminator method uses fewer toxic chemicals and lower amounts of radioactivity. The key principle of the Sanger method was the use of dideoxynucleotides triphosphates (ddNTPs) as DNA chain terminators.

The classical chain-termination or Sanger method requires a single-stranded DNA template, a DNA primer, a DNA polymerase, radioactively or fluorescently labeled nucleotides, and modified nucleotides that terminate DNA strand elongation. The DNA sample is divided into four separate sequencing reactions, containing the four standard deoxynucleotides (dATP, dGTP, dCTP and dTTP) and the DNA polymerase. To each reaction is added only one of the four dideoxynucleotides (ddATP, ddGTP, ddCTP, or ddTTP). These dideoxynucleotides are the chain-terminating nucleotides, lacking a 3'-OH group required for the formation of a phosphodiester bond between two nucleotides during DNA strand elongation. Incorporation of a dideoxynucleotide into the nascent (elongating) DNA strand therefore terminates DNA strand extension, resulting in various DNA fragments of varying length. The dideoxynucleotides are added at lower concentration than the standard deoxynucleotides to allow strand elongation sufficient for sequence analysis.

The newly synthesized and labeled DNA fragments are heat denatured, and separated by size (with a resolution of just one nucleotide) by gel electrophoresis on a denaturing polyacrylamide-urea gel. Each of the four DNA synthesis reactions is run in one of four individual lanes (lanes A, T, G, C); the DNA bands are then visualized by autoradiography or UV light, and the DNA sequence can be directly read off the X-ray film or gel image. In the image on the right, X-ray film was exposed to the gel, and the dark bands correspond to DNA fragments of different lengths. A dark band in a lane indicates a DNA fragment that is the result of chain termination after incorporation of a dideoxynucleotide (ddATP, ddGTP, ddCTP, or ddTTP). The terminal nucleotide base can be identified according to which dideoxynucleotide was added in the reaction giving that band. The relative positions of the different bands among the four lanes are then used to read (from bottom to top) the DNA sequence as indicated.

There are some technical variations of chain-termination sequencing. In one method, the DNA fragments are tagged with nucleotides containing radioactive phosphorus for radiolabelling. Alternatively, a primer labeled at the 5' end with a fluorescent dye is used for the tagging. Four separate reactions are still required, but DNA fragments with dye labels can be read using an optical system, facilitating faster and more economical analysis and automation. This approach is known as 'dye-primer sequencing'. The later development by L Hood and coworkers of fluorescently labeled ddNTPs and primers set the stage for automated, high-throughput DNA sequencing.

Sequence ladder by radioactive sequencing compared to fluorescent peaks (click to expand)

The different chain-termination methods have greatly simplified the amount of work and planning needed for DNA sequencing. For example, the chain-termination-based "Sequenase" kit from USB Biochemicals contains most of the reagents needed for sequencing, prealiquoted and ready to use. Some sequencing problems can occur with the Sanger Method, such as non-specific binding of the primer to the DNA, affecting accurate read out of the DNA sequence. In addition, secondary structures within the DNA template, or contaminating RNA randomly priming at the DNA template can also affect the fidelity of the obtained sequence. Other contaminants affecting the reaction may consist of extraneous DNA or inhibitors of the DNA polymerase.

Dye-terminator sequencing

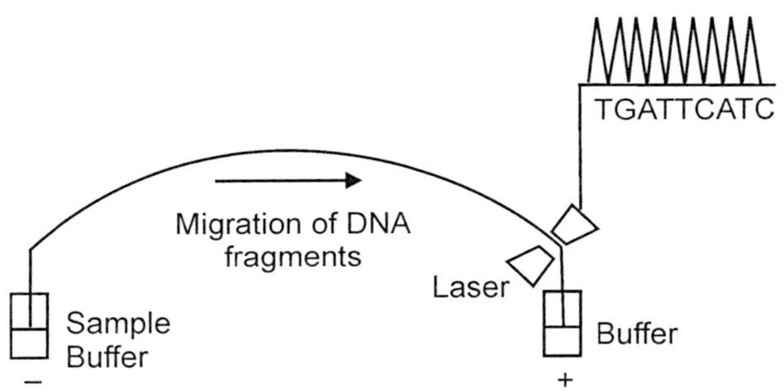

Fig. Capillary electrophoresis

An alternative to primer labelling is labelling of the chain terminators, a method commonly called 'dye-terminator sequencing'. The major advantage of this method is that the sequencing can be performed in a single reaction, rather than four reactions as in the labelled-primer method. In dye-terminator sequencing, each of the four dideoxynucleotide chain terminators is labelled with a different fluorescent dye, each fluorescing at a different wavelength. This method is attractive because of its greater expediency and speed and is now the mainstay in automated sequencing with computer controlled sequence analyzers. Its potential limitations include dye effects due to differences in the incorporation of the dye-labelled chain terminators into the DNA fragment, resulting in unequal peak heights and shapes in the electronic DNA sequence trace chromatogram after capillary electrophoresis. This problem has largely been overcome with the introduction of new DNA polymerase enzyme systems and dyes that minimize incorporation variability. The dye-terminator sequencing method, along with automated high-throughput DNA sequence analyzers, is now being used for the vast majority of sequencing projects, as it is both easier to perform and lower in cost than most previous sequencing methods.

Automation and Sample Preparation

Modern automated DNA sequencing instruments (DNA sequencers) can sequence up to 384 fluorescently labelled samples in a single batch (run) and perform as many as 24 runs a day. However, automated DNA sequencers carry out only DNA size separation by capillary electrophoresis, detection and recording of dye fluorescence, and data output as fluorescent peak trace chromatograms. Sequencing reactions by thermocycling, cleanup and re-suspension in a buffer solution before loading onto the sequencer are performed separately.

Large-scale Sequencing Strategies

Current methods can directly sequence only relatively short (300-1000 nucleotides long) DNA fragments in a single reaction. The main obstacle to sequencing DNA fragments above this size limit is insufficient power of separation for resolving large DNA fragments that differ in length by only one nucleotide. Limitations on ddNTP incorporation were largely solved by Tabor at Harvard Medical, Carl Fuller at USB biochemicals, and their coworkers.Large-scale sequencing aims at sequencing very long DNA fragments. Even relatively small bacterial genomes contain millions of nucleotides, and the human chromosome alone contains about 246 million bases. Therefore, some approaches consist of cutting (with restriction enzymes) or shearing (with mechanical forces) large DNA fragments into shorter DNA fragments. The fragmented DNA is cloned into a DNA vector, usually a bacterial plasmid, and amplified in Escherichia coli. The amplified DNA can then be purified from the bacterial cells (a disadvantage of bacterial clones for sequencing is that some DNA sequences may be inherently un-clonable in some or all available bacterial strains, due to deleterious effect of the cloned sequence on the host bacterium or other effects). These short DNA fragments purified from individual bacterial colonies are then individually and completely sequenced and assembled electronically into one long, contiguous sequence by identifying 100per cent-identical overlapping sequences between them (shotgun sequencing). This method does not require any pre-existing information about the sequence of the DNA and is often referred to as de novo sequencing. Gaps in the assembled sequence may be filled by Primer walking, often with sub-cloning steps (or transposon-based sequencing depending on the size of the remaining region to be sequenced). These strategies all involve taking many small reads of the DNA by one of the above methods and subsequently assembling them into a contiguous sequence. The different strategies have different tradeoffs in speed and accuracy; the shotgun method is the most practical for sequencing large genomes, but its assembly process is complex and potentially error-prone - particularly in the presence of sequence repeats.

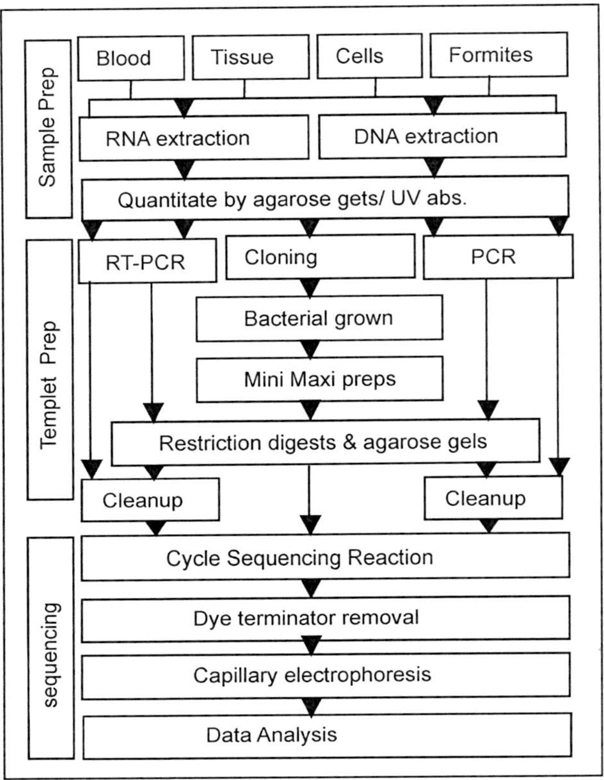

Fig. Resequencing steps. Sample prep: Extraction of nucleic acid. Template prep: Amplification and preparation of a small region of the target region.

The human genome is about 3 billion (3,000,000,000) bp long; if the average fragment length is 500 bases, it would take a minimum of six million (3 billion/ 500) to sequence the human genome (not allowing for overlap = 1-fold coverage). Keeping track of such a high number of sequences presents significant challenges, only held down by developing and coordinating several procedural and computational algorithms, such as efficient database development and management.

Resequencing or targeted sequencing is utilized for determining a change in DNA sequence from a "reference" sequence. It is often performed using PCR to amplify the region of interest (pre-existing DNA sequence is required to design the PCR primers). Resequencing uses three steps, extraction of DNA or RNA from biological tissue; amplification of the RNA or DNA (often by PCR); followed by sequencing. The resultant sequence is compared to a reference or a normal sample to detect mutations.

Next-generation technology

Because of strong demand for high-throughput sequencing technologies by publicly and privately funded research institutions and companies, many companies have research programmes to develop next-generation DNA-sequencing technologies, which may replace the current dye-terminator technology based on DNA separation by capillary electrophoresis.

The current market leader in DNA sequencing, the ABI division of Applera, has a new, rather baroque, technology, Solid, which combines elements of emulsion polymerization from Agencourt with an oligo ligation strategy and methods developed by Brenner et al at Lynx (now part of Solexa)As of Feb 2007, some of the other players include Perlegen, an Affymetrix subsidiary, which uses sequencing by hybridization, originally proposedby Drmanac et al.; Solexa, now owned by Illumina, which uses a Bridge amplification technology originally developed by Adams and Kron; Helicos, a private company in Cambridge, MA which emphasizes single molecule sequencing; 454 Life Sciences, a subsidiary of Curagen with funding from Roche, using the innovative pyrosequencing method developed by Ronaghi et al; a colony-based technique (from Mitra and Church at Harvard) developed by Agencourt and licensed to ABI, and probably others.

Other proposals include labeling the DNA polymerase, reading the sequence as a strand transits through nanopores, microscopy-based techniques capable of single molecule resolution, such as AFM or electron microscopy and many others. There is an extensive list of methods that have failed to become commercial successes, such as the blot-based genome sequencing method of Church and Gilbert, commercialized by Integrated Genetics, methods using tin isotope-tagged nucleotides and mass spec detectors; and the "bottom wiper" of Pohl commercialized by Hoefer.

MEDICAL GENETICS RESEARCH

Medical genetics research seeks to find and study the genetic causes of human diseases. When searching for an unknown gene that may be involved in a disease, researchers commonly use genetics linkage and genetic pedigree charts to find the location on the genome associated with the disease. At the population level, researchers take advantage of Mendelian randomization to look for locations in the genome that are associated with diseases, a technique especially useful for multigenic traits not clearly defined by a single gene. Once a candidate gene is found, further research is often done on the same gene (called an orthologous gene) in model organisms.

CHROMOSOMES, RECOMBINATION, AND LINKAGE

Genes are arranged linearly along the long chains of DNA sequence, called chromosomes. In bacteria, each cell has a single circular chromosome, while eukaryotic organisms (which includes plants and animals) have their DNA arranged in multiple linear chromosomes. These DNA strands are often extremely long; the largest human chromosome, for example, is about 140 million base pairs in length.The DNA of a chromosome is associated with structural proteins which organize, compact, and control access to the DNA, forming a material called chromatin; in eukaryotes chromatin is usually composed of nucleosomes, repeating units of DNA wound around a core of histone proteins.

While haploid organisms have only one copy of each chromosome, most animals and many plants are diploid, containing two copies of each chromosome and thus two copies of every gene. An exception exists in the sex chromosomes, specialized chromosomes many animals have evolved that play a role in determining the sex of the organism. On the other chromosomes (called autosomes), the two alleles for a gene are located on identical loci of sister chromatids.

The diploid nature of chromosomes allows for genes on different chromosomes to assort independently, thereby recombining to form new combinations of genes. Genes on the same chromosome would theoretically never recombine, however, were it not for the process of chromosomal crossover. During crossover, chromosomes exchange stretches of DNA, effectively shuffling the gene alleles between the chromosomes. This process of chromosomal crossover generally occurs during meiosis, a series of cell divisions that creates haploid germ cells which later combine with other germ cells to form child organisms.

The probability of chromosomal crossover occurring between two given points on the chromosome is related to the distance between them. For an arbitrarily long distance, the probability of crossover is high enough that the inheritance of the genes is effectively uncorrelated. For genes that are closer together, however, the lower probability of crossover means that the genes demonstrate genetic linkage. The linkages between a series of genes can be combined to form a linear linkage map that roughly describes the arrangement of the genes along the chromosome.

Epigenetic inheritance

Although DNA is the genetic material of life, there are some aspects of cells that are not encoded in DNA sequence (eg. chromatin and DNA modifications) that are inherited by daughter cells when cells divide. These features are called “epigenetic” — the prefix “epi-” means “on top of” or “in addition to”. Epigenetic features create differences between cells sharing the same genome, allowing cells to differentiate into different tissues in multicellular organisms. Although epigenetic features are generally dynamic over the development of organisms, some, like the phenomenon of paramutation, have multigenerational inheritance.

MODIFYING AND REGULATOR GENES

There are two classes of genes that can have an effect on how other genes function. They are called modifying genes and regulator genes.

Modifying genes alter how certain other genes are expressed in the phenotype. For instance, there is a dominant cataract gene which will produce varying degrees of vision impairment depending on the presence of a specific allele for a companion modifying gene. However, cataracts also can be promoted by diabetes and common environmental factors such as excessive ultraviolet radiation.

Regulator genes can either initiate or block the expression of other genes. They control the production of a variety of chemicals in plants and animals. For instance, the time of production of specific proteins that will be new structural parts of our bodies can be controlled by such regulator genes. Shortly after conception, regulator genes work as master switches orchestrating the timely development of our body parts. They are also responsible for changes that occur in our bodies as we grow older. In other words, they control the maturation and aging processes. Regulator genes that are involved in subdividing an embryo into what will become the major body parts of an individual are also referred to as homeotic or homeobox genes. They are responsible for setting generalized cells on the path to become a head, torso, arms, legs, etc.

Incomplete Penetrance

Some genes are incompletely penetrant. That is to say, their effect does not normally occur unless certain environmental factors are present. For example, you may inherit a gene for diabetes but never get the disease unless you become greatly overweight, persistently stressed psychologically, or do not get enough sleep on a regular basis. Similarly, the gene that causes the chronic disease known as multiple sclerosis may be triggered by the Epstein-Barr virus.

Sex Related Genetic Effects

There are three categories of genes that may have different effects depending on an individual's gender. These are referred to as:

1. Sex-limited genes
2. Sex-controlled genes
3. Genome imprinting

Sex-limited genes are ones that are inherited by both men and women but are normally only expressed in the phenotype of one of them. The heavy male beard is an example. While women have facial hair it is most often very fine and comparatively sparse. In contrast, sex-controlled genes are expressed in both sexes but differently. An example of this is gout, a disease that causes painfully inflamed joints. If the gene is present, men are nearly eight times more likely than women to have severe symptoms. Some genes are known to have a different effect depending on the gender of the parent from whom they are inherited. This phenomenon is referred to as genome imprinting or genetic imprinting. Apparently, diabetes, psoriasis, and some rare genetically inherited diseases, such as a form of mental retardation known as Angelman syndrome, can follow this inheritance pattern.

Pleiotropy

A single gene may be responsible for a variety of traits. This is called pleiotropy. The complex of symptoms that are collectively referred to as sickle-cell trait, or sickle-cell anemia, is an example. A single gene results in irregularly shaped red blood cells that painfully block blood vessels, cause poor overall physical development, as well as related heart, lung, kidney, and eye problems. Another pleiotropic trait is albinism. The gene for this trait not only results in a deficiency of skin, hair, and eye pigmentation but also causes defects in vision.

Stuttering Alleles

Lastly, it is now known that some genetically inherited diseases have more severe symptoms each succeeding generation due to segments of the defective genes being doubled in their transmission to children (as illustrated below). These are referred to as stuttering alleles or unstable alleles. Examples of this phenomenon are Huntington's disease, fragile-X syndrome, and the myotonic form of muscular dystrophy.

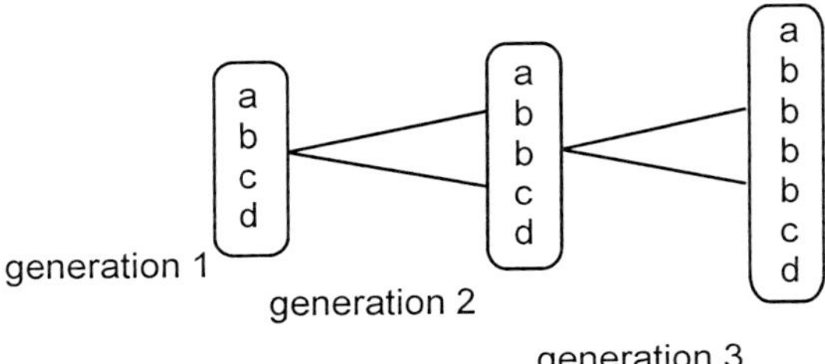

Fig. Unstable allele doubling each generation

Mendel believed that all units of inheritance are passed on to offspring unchanged. Unstable alleles are an important exception to this rule.

Environmental Influences

The phenotype of an individual is not only the result of inheriting a particular set of parental genes.The specific environmental characteristics of the uterus in which a fertilized egg is implanted and the health of the mother can have major impacts on the phenotype of the future child. For instance, oxygen deprivation or inappropriate hormone levels can cause lifelong, devastating effects. Likewise, accidents, poor nutrition, and other environmental influences throughout life can alter an individual's phenotype. Geneticists study identical or monozygotic twins to determine which traits are inherited and which ones were acquired following conception. Since monozygotic twins come from the same zygote, they are essentially identical in their genetic makeup. If there are any differences in their phenotypes, the environment is virtually always responsible. Such differences show up in basic capabilities such as handedness, which had been assumed to be entirely genetically determined. In rare instances, one monozygotic twin will be clearly right-handed while the other will be left-handed. This suggests that there may be both genetic and environmental influences in the development of this trait.

DOMINANT AND RECESSIVE CHARACTERS RESULTING FROM SUBSTITUTION OF CHROMOSOMES

The daughters produced in the experiment of contain X-chromosomes from two different sources one from the normal-eyed mother, one from the bar-eyed father.The one from the mother gives when by itself a normal eye, the one from the father a bar-eye.

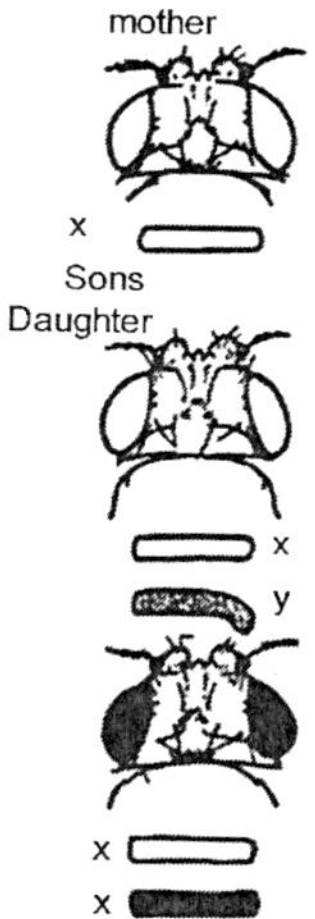

Fig. Inheritance of a recessive characteristic that depends on a particular type of X-chromosome: white eyes. The mother has white eyes, the father red eyes (represented in black]. The X-chromosomes of each individual are represented below the Fig of the head, the mothers chromosomes in outline, the father's in black. The daughters, receiving one X-chromosome from each parent, have red eyes, like the father. The sons, receiving an X-chromosome from the mother only, have white eyes, like the mother. The white eyes appear wherever the maternal type ofX (outline) is the only type ofX that is present.

As the daughters have bar-eyes, it is evident that the effect of the chromosome from the father prevails over the effect of that from the mother. It is the custom to call the character that thus prevails and is manifested, when two chromosomes of different tendency are present, a dominant character. Thus bar-eye is dominant, as compared with normal eye.Other X-chromosomes, in other individuals of Drosophila, are found to produce still other effects. Some are defective or abnormal, but instead of producing bar-eyes they produce other abnormalities. And most of them behave in a way that differs from the behaviour of bar-eye. An example will make clear this behaviour. In certain cases X-chromosomes produce eyes that are white instead of red. Normally in the fruit-fly the eyes are red. But sometimes there are found individuals whose eyes are white. When these are mated together, they produce offspring all of whose eyes are white. So the white eyes are hereditary.But what happens if one mates together red-eyed and white-eyed parents? What will be the programmes of the eyes in the offspring? Mate first a white-eyed mother with a red-eyed father.

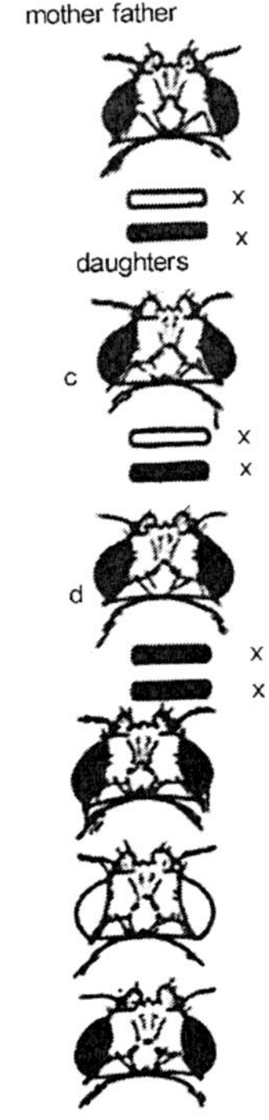

Fig. Course of inheritance of a recessive character (white eyes) that is dependent on a defective X. The mother has one defective X- chromosome (shown in outline), derived from the white-eyed mother of Fig. ii, her other X-chromosome being normal; she therefore has normal red eyes. The father has a normal X-chromosome (black), and normal red eyes. Of the four types of offspring shown (a, b, c, d), only one (c), representing half the sons, have the white eyes, this being the consequence of their receiving the mother's defective X-chromosome. All the daughters (a and b) and the other half of the sons (d) have normal redeyes, since they contain at least one of the normal X-chromosomes.

In this case the Fig represents the mother's X-chromosome in white (since these are the abnormal individuals), while the father's are shown in black.Thus the descendants of the two kinds can be traced in later generations.The sons get an X-chromosome from the mother only. And they all have white eyes, like the mother. The daughters get one X from the mother, one from the father. And they all have red eyes, like the father.

So the white eyes go with the mother's X-chromosome. But if one of the father's X-chromosomes is likewise present, then the white eyes are not produced; such daughters have red eyes. So the red eyes are dominant. The white eyes, since they do not appear if a normal X-chromosome from the father is present along with the abnormal one from the mother, are said to be recessive. White eyes appear only in individuals in which the abnormal X-chromosomes are the only kind of Xpresent.

All this can be tested in a great number of ways. Mate, for example, one of these daughters, having the two kinds of chro-mosomes, with a normal red-eyed male, having a normal (black) X. Now half of the offspring get from the mother a defective X (white), the other half get from her a normal X (black). This is true both for the daughters and the sons. But the daughters get also a normal X from the redeyed father. So they all have normal red eyes. But the sons do not get an X from the father; the only X they have is the one from the mother. And those that get the normal X from the mother have red eyes, while those that get the defective X have white eyes.

That is, there is now produced a family in which all the daughters have red eyes, while half the sons have red eyes and the other half have white eyes. In a family of two hundred, there are a hundred red-eyed daughters, fifty red-eyed sons, and fifty white-eyed sons.

The white eyes are present in all cases where the efective chromosome from the grandfather is the only kind present. There are many ways to test this, and it is always verified. For example, if we cross one of the red-eyed mothers that has one defective X-chromosome, with a white-eyed father that also has the defective X-chromosome then we find that half the sons, and also half the daughters, have white eyes, the other half red eyes. And this is exactly the result given if the white eyes appear only where the defective X is the only kind present.Since such characters appear in all males that have the defective X (since they have no other), while in females they appear only if both the X's are defective, they are much commoner in males than in females, and are, hence, often called sex-linked characters. But they appear in females as well as in males, provided that both the X's are defective. Many Types of X-Chromosomes exist.

Thus it is clear, up to this point, that there are among the different individuals of the fruit-fly at least three kinds of X-chromosomes, with different effects. Some are normal, producing normal red eyes. Others have become defective in such a way that they produce bar-eyes. Others have become defective in such a way that they produce white eyes.These are not the only kinds of changed X-chromosomes.

Many different sorts of peculiarities have been found that thus follow the distribution of a particular type of X-chromosome, being manifested in the body of the individual when that type of X is the only type present. Up to 1925, a hundred and thirty diverse types of X-chromosomes had been detected scattered among the different individuals in the fruit-fly each type giving rise to some special peculiarity. Many are known also in other organisms. In man a considerable number are known, giving rise mainly to defects or diseases.Among the characteristics dependent on changed X-chromosomes in man are some types of colourblindness, also a blood disease known as haemophilia. They follow exactly the distribution of a particular type of X-chromosome. This, ofcourse, shows that in man too the male has but a single X chromosome, while the female has two, for this is the only way in which that method of inheritance can be brought about.

Thus it is established that when one X-chromosome is sub stituted for another, this commonly makes a difference in the characteristics of the individuals. Since there are many types of X-chromosomes distributed among the individuals of a single species, such substitutions can be made in many different ways, giving many different types of results. In the fruitfly, some X-chromosomes produce red eyes, others produce white eyes, others buff coloured, or eosin coloured, or various other coloured eyes. Some X-chromosomes produce long wings, others short wings and so one could list a hundred or more diverse kinds of effects of diverse X-chromosomes. Similar long lists could be made for certain other organisms, and a considerable number of diverse types of X-chromosomes are known even in man.The other chromosomes present somewhat more complex conditions than do the X-chromosomes, because they are always in pairs, while in some individuals (males) X is single. This makes detection of the effects of substituting one chromosome for another more difficult. But the same type of experimentation illustrated above for the X-chromosomes demonstrates that the other chromosomes also influence the characteristics of organisms. These matters are taken up in detail in later chapters. It is clear therefore that the chromosomes are indeed 'materials of heredity'. That is, they influence the characteristics of the individuals that contain them, for these characteristics are altered when one chromosome is substituted for another. We turn next to a study of these materials.

The relations between the materials of heredity and the characteristics of the organism have been far more thoroughly investigated in the fruit-fly, Drosophila melanogaster, than in any other organism, so that much of the detailed knowledge of these matters is based upon this animal. Most of the fundamental work on Drosophila has been done by T. H. Morgan and his associates,G. B. Bridges and A. H. Sturtevant, though in recent years many investigators in all parts of the world have been at work on the genetics of this organism, adding greatly to our knowledge. When Mendel studied peas, one of the phenotypes showed complete dominance over the other one. If we look at pea height, and denote the gene for short as s and the gene for tall as S, then as every plant has two sets chromosomes each has two genes at this locus.So there are three possibilities: SS, Ss, ss (order doesn't matter) In this case S was fully dominant over s, so Ss individuals were phenotypically identical to SS individuals. Only ss pea plants were short. The S gene would be said to be Dominant.While the s gene is said to be Recessive.

GENE IS A SEGMENT OF NUCLEIC ACID

A gene is a segment of nucleic acid that contains the information necessary to produce a functional RNA product in a controlled manner. They contain regulatory regions dictating under what conditions this product is made, transcribed regions dictating the sequence of the RNA product, and/or other functional sequence regions. Genes interact with each other and the environment to determine the physical development and behaviour of organisms, they are units of inheritance.

In cells, genes consist of a long strand of DNA that contains a promoter, which controls the activity of a gene, and a coding sequence, which determines what the gene produces. When a gene is active, the coding sequence is copied in a process called transcription, producing an RNA copy of the gene's information. This RNA can then direct the synthesis of proteins via the genetic code. However, RNAs can also be used directly, for example as part of the ribosome.These molecules resulting from gene expression, whether RNA or protein, are known as gene products.

Most genes contain non-coding regions that do not code for the gene products, but regulate gene expression. The genes of eukaryotic organisms can contain non-coding regions called introns that are removed from the messenger RNA in a process known as splicing. The regions that actually encode the gene product, which can be much smaller than the introns, are known as exons. One single gene can lead to the synthesis of multiple proteins through the different arrangements of exons produced by alternative splicings.

The total complement of genes in an organism or cell is known as its genome. The genome size of an organism is loosely dependent on its complexity; prokaryotes such as bacteria and archaea have generally smaller genomes, both in number of base pairs and number of genes, than even single-celled eukaryotes. However, the largest known genome belongs to the single-celled amoeba Amoeba dubia, with over 6 billion base pairs. The estimated number of genes in the human genome has been repeatedly revised downward since the completion of the Human Genome Project; current estimates place the human genome at just under 3 billion base pairs and about 20,000-25,000 genes.. A recent Science article gives a final number of 20,488, with perhaps 100 more yet to be discovered . The gene density of a genome is a measure of the number of genes per million base pairs (called a megabase, Mb); prokaryotic genomes have much higher gene densities than eukaryotes. The gene density of the human genome is roughly 12-15 genes/Mb.

CONCLUSION

How a living organism is built and function determined and govern by genes. By understanding how genes work:

- It would give the possibility to cure genetic illnesses.
- It would give the possibility to determine the organism's features and behaviors.
- It would give the possibility to create a new organism.

The relationships between genes and features are very complex. It may not be possible to change one and only one feature of an organism by changing one or set of genes(based on current knowledge). A feature may be changed, but probably not alone, only with other changes. The other changes could be small and/or ignorable however. Some changes are gradual, while changes in gene expression can result in rapid transformations in the physiological state of an organism.

Genetics is the study of the function and behaviour of genes. Genes are bits of biochemical instructions found inside the cells of every organism from bacteria to humans. Offspring receive a mixture of genetic information from both parents. This process contributes to the great variation of traits that we see in nature, such as the colour of a flower's petals, the markings on a butterfly's wings, or such human behavioral traits as personality or musical talent. Geneticists seek to understand how the information encoded in genes is used and controlled by cells and how it is transmitted from one generation to the next. Geneticists also study how tiny variations in genes can disrupt an organism's development or cause disease. Increasingly, modern genetics involves genetic engineering, a technique used by scientists to manipulate genes. Genetic engineering has produced many advances in medicine and industry, but the potential for abuse of this technique has also presented society with many ethical and legal controversies.

Genetic information is encoded and transmitted from generation to generation in deoxyribonucleic acid (DNA). DNA is a coiled molecule organized into structures called chromosomes within cells. Segments along the length of a DNA molecule form genes. Genes direct the synthesis of proteins, the molecular laborers that carry out all lifesupporting activities in the cell. Although all humans share the same set of genes, individuals can inherit different forms of a given gene, making each person genetically unique. Since the earliest days of plant and animal domestication, around 10,000 years ago, humans have understood that characteristic traits of parents could be transmitted to their offspring. The first to speculate about how this process worked were Greek scholars around the 4th century bc, who promoted theories based on conjecture or superstition. Some of these theories remained in favour for several centuries. The scientific study of genetics did not begin until the late 19th century. In experiments with garden peas, Austrian monk Gregor Mendel described the patterns of inheritance, observing that traits were inherited as separate units. These units are now known as genes. Mendel's work formed the foundation for later scientific achievements that heralded the era of modern genetics.

Most of the discoveries about genes and DNA happened after Charles Darwin 1859 famous book on The Origin of Species by Means of Natural Selection, or the preservation of favoured races in the struggle for life.

Genetics is the science of heredity and variation in living organisms. Knowledge that desired characteristics were inherited has been implicitly used since prehistoric times for improving crop plants and animals through selective breeding. However, the modern science of genetics, which seeks to understand the mechanisms of inheritance, only began with the work of Gregor Mendel in the mid-1800s.

Mendel observed that inheritance is fundamentally a discrete process with specific traits that are inherited in an independant manner. These basic units of inheritance is now known as "genes". In the cells of organisms, genes exist physically in the structure of the molecule DNA and the information genes contain is used to create and control the components of cells. Although genetics plays a large role in determining the appearance and behaviour of organisms, it is the interaction of genetics with the environment an organism experiences that determines the ultimate outcome. For example, while genes play a role in determining a person's height, the nutrition and health that person experiences in childhood also have a large effect.

Classical genetics consists of the techniques and methodologies of genetics that predate the advent of molecular biology. A key discovery of classical genetics in eukaryotes was genetic linkage. The observation that some genes do not segregate independently at meiosis, broke the laws of Mendelian inheritance, and provided science with a way to map characteristics to a location on the chromosomes. Linkage maps are still used today, especially in breeding for plant improvement.

After the discovery of the genetic code and such tools of cloning as restriction enzymes, the avenues of investigation open to geneticists were greatly broadened. Some classical genetic ideas have been supplanted with the mechanistic understanding brought by molecular discoveries, but many remain intact and in use. Classical genetics is often contrasted with reverse genetics, and aspects of molecular biology are sometimes referred to as molecular genetics.

2

The Genetic System

CERTAIN TERMS DEFINED

In connection with the above account of the history of the materials of heredity, it will be well to observe the meaning of certain common terms, whose use is very convenient; they will be employed in later pages.

Gametes: The mature germ cells (ova and sperm) are known as gametes. In the gametes the chromosomes are single instead of in pairs: that is, the gametes contain but n chromosomes, in place of zn. Cells having thus but one set of chromosomes are said to be haploid, while the usual cells, with chromosomes in pairs, are diploid.The entire process of producing gametes is known as gametogenesis. The later processes, by which the four mature germ cells are produced, are spoken of as maturation. The chromosomal changes during maturation, resulting in reduction in the number of chromosomes from 2n to n_y are known as meiosis.

Zygote: The fertilized egg, formed by the union of two gametes, is the zygote. This term is often applied to the individuals developed from the zygote. The zygote is diploid: its chromosomes are in pairs, their total number being 2fl. The union of the gametes to form the zygote is spoken of as the fertilization of the ovum.Heterogametic: The individuals of one sex (the male in the types described above) form two kinds of gametes, one containing an X-chromosome, the other not. Such individuals are said to be heterogametic (or sometimes the term digametic is employed in the same sense).

Autosomes and Sex Chromosomes: The chromosomes X and Y, since they play a prominent part in the production of sex, are known as sex chromosomes. All the other chromosomes are known as autosomes. Thus Drosophila melanogaster has one pair of sex chromosomes and three pairs of autosomes.

THE CHROMOSOMES BETWEEN THE CELL DIVISIONS, WHEN THE CELLS ARE NOT DIVIDING

1. At the time of cell division the chromosomes are in the condensed condition.

2. After completion of cell division, each chromosome absorbs fluid from the cytoplasm, grows, enlarges, and becomes a small vesicle. This is a slow, gradual process, that can be seen under the microscope.
3. The vesicles become so large that their boundaries touch, and are pressed together, the boundaries becoming very indistinct or completely disappearing.
4. The entire set of zn (or 2fl-i) vesicles derived from the chromosomes constitutes the nucleus (what is called the 'resting nucleus').
5. In some organisms the entire set of zn chromosome vesicles can be dimly seen and counted, even in this resting stage. In others this cannot be done.
6. As any cell gets ready to divide again, a new, small chromosome is laid down in each of the vesicles formed by the previous chromosomes.
7. When these new chromosomes first appear, they are usually irregular granules, which gather into a thread, having thickenings at intervals. In this condition they are called the skein or spireme.
8. These threads gather together and condense into the chromosomes of the type seen at cell division.
9. The remainder of each large chromosomal vesicle dissolves in the cytoplasm of the cell. Here it doubtless produces chemical changes of importance.
   Thus after every cell division the chromosome takes up a quantity of fluid from the cytoplasm, and later gives it off again, doubtless in modified form. This is probably one of the most important processes in development. It is presumably in this way that the chromosomes influence the development and the characteristics of the organism. They take up parts of the cytoplasm, change it, and give it off again. In this way are probably made the different kinds of tissues: muscles, nerves, bones, and the like. The chromosomes are thus continually active in the process of manufacturing the body (a matter to which we return later).
10. While thus in vesicular form, and active in the physiological processes of the cell, the chromosomes as such have in most cases disappeared; they are no longer visible (though in exceptional cases they can still be detected). When later they reappear they show definite relations to the chromosomes that were visible before disappearance, in the following particulars:
    a. The same number reappears. If the number is changed experimentally, as has often been done, it is the altered number that reappears.

b. The chromosomes reappear in the same set of diverse forms and sizes as before. This is very striking in organisms like Drosophila in which the different pairs of chromosomes differ much. In hybrids, often the maternal and paternal sets differ greatly in size and form; each set reappears in its own type.
c. The chromosomes reappear in the original grouping and arrangement. When the chromosomes that disappeared were in pairs, as in Drosophila they reappear paired as before. Often at disappearance the maternal and paternal sets are in two separate groups they reappear in these same groups. Often two particular chromosomes that are close together or intertwined reappear in this same condition.

Thus, even in cases in which the chromosomal vesicles are not separately detectable during the 'resting stage', it is obvious that each chromosome reproduces itself, in the same form, size, and position.

n. All this is repeated at every cell division (though the arrangement or relative position of the chromosomes gradually changes in the course of many cell divisions). Thus finally each cell of the body contains the complete set of 2/ 2 chromosomes, half from the mother, half from the father.

THE GENETIC SYSTEM IN THE FIRST STAGE OF THE INDIVIDUAL; THE FERTILIZED EGG

1. As the new individual begins its separate life in the form of a fertilized egg, it contains, embedded in the cytoplasm of the cell, two nuclei, each with a set of chromosomes.
2. One of these nuclei is from the mother (the 'female pro- nucleus'), the other from the father (the 'male pronucleus').
3. The two pronuclei contain the same number of chromosomes (or that from the father contains one less; see the following). The number contained differs in different organisms; it will be convenient to call the number n. Thus each of the two pronuclei has a set of n (or n-i) chromosomes, and the entire egg has two sets, making zn (or in the male, 2w-i) chromosomes.

   The number «, present in each of the two pronuclei, is known as the haploid number. The number 2W (or 2«-i), present in the two together, is known as the diploid number. In different organisms the haploid number varies from i to more than 100.
4. In some of the fertilized eggs (those that will develop into males), in some organisms, the pronucleus from the mother contains one more chromosome than that from the father. This additional chromosome is known as X. In the eggs that are to produce females (in these same organisms) the two pronuclei contain the same number of chromosomes, each having an X.

5. In organisms in which the number n is the same for both pronuclei, frequently in the eggs that are to produce males one chromosome (X) in the maternal pronucleus is larger than the corresponding chromosome (Y) in the paternal pronucleus.
6. In some organisms the different chromosomes in each single pronucleus differ much in size and shape.
7. But the set of chromosomes in one of the two pronuclei is like the set in the other, as to size and form of the chromosomes, except in the case of the X and Y chromosomes above mentioned.
8. The two pronuclei are at first separate; they approach one another, come in contact, their membranes dissolve, and the two groups of chromosomes come closer together, forming a single group of 2« chromosomes (or 2«-i). This is the diploid group.
9. Later this group becomes surrounded by a single membrane, constituting thus the single nucleus of the fertilized egg, containing the diploid number of chromosomes.
10. Thus in the single nucleus of the fertilized egg, half of the chromosomes are from the mother, half (or one less) from the father.
11. Sooner or later the corresponding chromosomes of the paternal and maternal sets mate or conjugate, forming a pair. In some animals, as in the flies, this pairing of the chromosomes takes place soon after the union of the two pronuclei (in the 2-cell stage).
12. Thus the nucleus of the fertilized egg contains n pairs of chromosomes, one member of each pair paternal in origin, the other maternal.
13. In some organisms (the flies and others), the two chromosomes of a pair (maternal and paternal) remain paired throughout the life of the individual, and in all his cells.
14. In other species the two members do not pair till later (when germ cells are formed). But finally they mate, in practically all organisms. We shall later examine what happens in mating of the chromosomes.
15. Even in species in which the maternal and paternal members are not at all times side by side, it is often possible to determine under the microscope which two chromosomes form a pair, since the pairs are of different sizes and forms, while the two members of the same pair are of the same size and form.

    The paired condition of the chromosomes turns out to be of great physiological importance.

THE PRODUCTION OF GERM CELLS, IN RELATION TO THE MATERIALS OF HEREDITY

Having followed the main features in the behaviour of the chromosomes in forming the body of the individual, we next examine what happens when the individual reproduces: the processes in forming germ cells and in producing offspring. In dealing with heredity, it is particularly important to have these correctly in mind.

1. The individual begins as a single cell, the fertilized egg.
2. It divides into many cells, some of which produce the body, while others produce the germ cells (G).
3. At an early stage in development there is usually among the numerous cells one that is later to produce all the germ cells; this is the primordial germ cell.
4. The primordial germ cell in most organisms has the same set of chromosomes and genes as have the body cells. (There are exceptions to this in certain organisms.) That is, the primordial germ cell has n pairs of chromosomes, making 2« (or 2fl-i) chromosomes. Of these, n are maternal in origin, while n (or n-i) are paternal in origin.

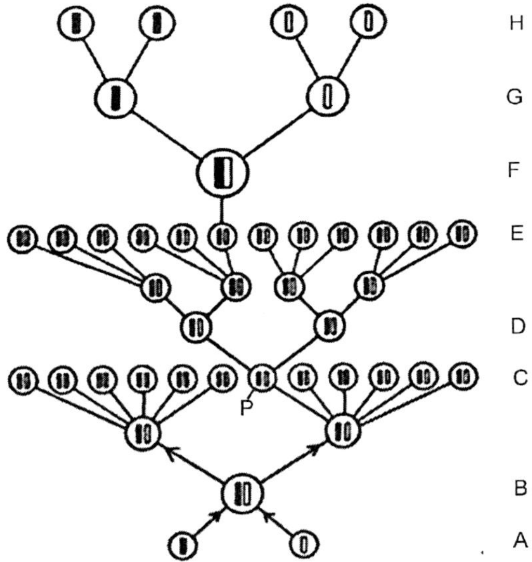

Fig. Diagram to illustrate the relation of body cells and germ cells) and the origin of both from the fertilized egg. Only a single pair of chromosomes is represented in each cell. At A are the two gametes (sperm and ovum) which unite to produce the fertilized egg B; each gamete contains one chromosome of the pair. The fertilized egg B divides, producing body cells C, among which is the primordial germ cell P. The primordial germ cell divides to produce many immature germ cells (E). The immature germ cell enlarges, the two chromosomes of a pair conjugating (F), then divides into four germ cells (G, H), each containing but one chromosome of the pair. (In the female three of the four germ cells (H) thus formed are not functional.}

5. The primordial germ cell divides many times, to produce many immature germ cells. These still contain the full two sets, n pairs, of

chromosomes. The immature germ cells in the male are the spermatogonia, while those in the female are oogonia.

6. There is usually a period of youth of the organism, in which the immature germ cells remain at rest, not dividing actively.
7. As the individual becomes mature, there is a renewed multiplication of the immature germ cells, the chromosomes remaining as before. Next follows a series of processes by which the final germ cells are produced: in their totality they are known as maturation. They include important changes in the chromosomes; these changes taken together are known as meiosis. Maturation and meiosis include the following:
8. Growth Stage: Some of the immature germ cells now grow, becoming much larger. After becoming large, the spermatogonia are called primary spermatocytes, while the oogonia are called primary oocytes.
9. Each primary spermatocyte and oocyte is later to divide twice, producing four mature germ cells.
   While the growth and division are occurring, important processes occur in the chromosomes, as follows:
10. The chromosomes pair if they have not already done so; that is, the corresponding paternal and maternal chromosomes place themselves side by side close together.
11. While pairing, the chromosomes are in the form of very long slender threads, which have thickenings at intervals. There are corresponding thickenings side by side in the two threads.
12. The two chromosomal threads (maternal and paternal) forming a pair become intimately united. This process is known as the conjugation, or synapsis, of the chromosomes. (During conjugation the two chromosomes may exchange parts, by 'crossing-over, as will be described later.)
    This intimate union of the maternal and paternal chromosomes appears to be the ultimate act in the union of the sexes.
13. Next occur certain processes, the general result of which is simple, although the processes are complex. The final result is that which would be produced if the immature germ cell now divided but once into two final germ cells and, in so doing, the two chromosomes of each pair separated, one going to one of the two germ cells, one to the other. Thus each mature germ cell contains finally one chromosome (the maternal one or the paternal one) from each pair. The total number of chromosomes in the germ cell is now one-half the number originally present in the immature germ cell; it is n in place of 2». This is the process known as reduction of the number of chromosomes.The actual processes are more complex, and are somewhat confusing. In following them, the final result just set forth should be kept in mind. They are as follows:

14. During conjugation the two long chromosomal threads shorten and thicken.
15. And while this is occurring, each of the two splits into two, so that there are now four partly united threads, two maternal, two paternal, in origin.
16. The shortening and thickening continues till the four threads have gathered into four closely united lobes. These are known as 'groups of four', or 'tetrads'.
    Since each pair of chromosomes becomes one tetrad, there are n tetrads. In each tetrad, two lobes represent a maternal chromosome, while two represent a paternal chromosome.
    Some males have one chromosome (X) without a mate; this does not conjugate, but otherwise goes through the same process as the others, forming a two-lobed instead of a four- lobed body.
17. Now the large cell containing the n tetrads divides quickly, twice in succession ('maturation divisions'), thus producing four final germ cells (sperms or ova) .
18. In these two maturation divisions, each tetrad divides twice, in such a way that the four lobes separate into the four different germ cells formed.
19. Thus two of the four germ cells receive from each tetrad a maternal lobe only, while two receive a paternal lobe only.
20. These lobes transform each into a chromosome like that from which it was derived.
21. Thus each final germ receives either a maternal or a paternal chromosome from each of the n pairs that were present in the immature germ cells. Each therefore contains but n single chromosomes, in place of the n pairs (or zn single chromosomes) that were present in the parental cells. This is the process of reduction of the number of chromosomes, from 2n to n.
    As set forth in paragraph 13 above, the net result is the same as would be produced by the division of each immature germ cell into two instead of four, one chromosome from each pair passing into each of the two germ cells produced. For many practical purposes, one can think of germ-cell formation as occurring in this way.
22. In the female, all the germ cells produced have the same number of chromosomes (fl), since each receives an X chromosome. In some species, in which the male has an odd number (2«-i) of chromosomes, half the germ cells receive n chromosomes, the other half receiving n-i.
23. In the male, all the four germ cells produced by the two maturation divisions are functional; they transform into gametes, known in the males as sperms.

24. In the female, three of the four cells formed by the two maturation divisions are small and without function; they are known as the polar bodies. The fourth is large, and is the functional gamete or ovum, which may be fertilized and may develop into a new individual.
25. A sperm with n (or n-i) chromosomes unites with an ovum containing n chromosomes, giving again a fertilized egg, or zygote, containing n pairs of chromosomes.
26. Different germ cells from the same parent, and different fertilized eggs produced by the same two parents, receive different combinations of chromosomes. This results from the following facts:
    a. Each parent carries n pairs of chromosomes, one chromosome of each pair being maternal, the other paternal, in origin.
    b. As we saw in the preceding chapter, the paternal and maternal chromosomes of a pair often differ in their effects. We may therefore designate them differently, calling one of them A, the other a, so that the pair is Aa.
    c. The chromosomes of the different pairs are often diverse in size and form; and, as will be shown later, they are also diverse in their functions.

       Therefore we may designate the chromosomes of the different pairs by different letters. Thus the maternal chromosomes of the different pairs may be represented by the capital letters A, B, C, and so on, while the paternal chromosomes of the same pairs may be represented by the small letters a, b, c, and so on. The series of pairs in the cells of the parent are thus Aa, Bb, Cc, and so on, for n pairs.
    d. As before seen, each final germ cell receives one chromosome from each pair. It may receive A or a from the pair Aa; B or b from the pair Bb, and so on.
    e. The different pairs are independent in their distribution to the germ cells. Thus from the two pairs Aa and Bb a given germ cell may receive any of the four combinations AB, Ab, aB, and ab; and similarly for other pairs. Any one combination occurs as frequently as any other.
    f. Thus with relation to one pair, as Aa, two types of germ cells are possible, A and a; similarly two types are possible from the pair Bb. Either type from Aa may be combined with either type from Bb, giving four diverse types from the two together, as illustrated above. Each additional pair multiplies the number of possible combinations by 2.
    g. Thus if the parent has n pairs of chromosomes, and the chromosomes of maternal origin differ in effect from those of

paternal origin, the number of different types of germ cells producible as a result of this situation is 2^n, these having diverse combinations of chromosomes in the different germ cells produced by that parent.

Thus in Drosophila melanogaster, which has four pairs of chromosomes, the number of diverse types of germ cells that may be thus formed is 2, or 16. If the four pairs are Aa, Bb, Cc, and Dd, then in the different germ cells there are such combinations as ABCD, AbCD, aBcd, aBcD, and so on. An organism with 10 pairs of chromosomes similarly produces 1024 diverse types of germ cells. In man, having 24 pairs of chromosomes, the number of diverse types producible is 2, or more than a million.

(As will be shown later, the number of diverse types of germ cells producible is greatly increased beyond these Figs by the fact that the two chromosomes of a pair may during conjugation exchange parts in various proportions.

27. Thus any single individual produces many diverse types of germ cells, containing diverse combinations of its chromosomes. This is true both for the sperms and the ova.

28. In fertilization, sperms and ova having the various diverse combinations of chromosomes unite at random. Thus the many fertilized eggs, and the offspring developed from them, bear different combinations of chromosomes. In consequence, as will be shown later, they may develop differently, giving many diverse types of individuals. The numbers of different types thus producible by g, given pair of parents is dealt with in later chapters.

THE CHROMOSOMES IN THE CELL DIVISIONS THAT PRODUCE THE BODY

1. The egg divides into 2 cells, these into 4, and so on, to 8, 16, and the like, till a great number of cells are formed, making up the body of the adult organism.
2. At each cell division, each of the 2« chromosomes divides into 2, by splitting lengthwise.
3. One half of each chromosome passes into each of the two cells formed by division.
4. There, each half grows to form a complete chromosome. Thus the chromosomes reproduce by fission, like Protozoa.
5. Consequently each cell formed contains the full set of 2« (or 2«-i) chromosomes, half of them maternal, half paternal.

Ultimately therefore every cell of the body contains the set of 2w (or 2«-i) chromosomes, half maternal, half paternal, in origin. (There are some exceptions to this statement.)

IN THE END, WE CAN SAY

In earlier chapter it is described the discovery of the materials of heredity, and the methods by which it was proved that they do indeed affect the characteristics of organisms. These materials, it was shown, lie in the chromosomes. And it was seen that removing a chromosome, or substituting one for another, has great and varied effects on individuals.

The study of genetics is very largely therefore the study of these materials, and particularly the study of their physiological action, in reproduction and heredity. As a foundation for understanding their action, it is necessary to have clear ideas of the main features of their location and history, from the formation of the new individual, through its development to an adult, and thence to the formation of a new generation. This forms the subject of the present chapter. Though here heredity, variation and characteristics are not explicitly considered, it will be found that for their understanding, as taken up in later chapters, knowledge of the matters set forth in this chapter is essential.

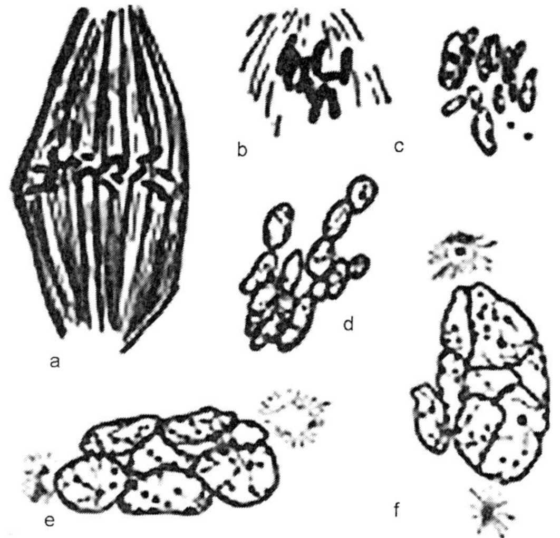

Fig. Transformation of the condensed chromosomes (a) by absorption of fluid into vesicles which slowly enlarge (b, c, d, e,f), finally constituting the nucleus. The Fig show the process in the cells of the egg of a fish, Fundulus.

The materials of heredity are gathered into a system of structures, which we may call the Genetic System. This system lies within the cells, and is as well defined in its constitu-tion and functions as is the nervous or muscular system. It includes the chromosomes; whether it includes also other parts lying outside the chromosomes will be considered later. For the present it is the chromosomes that will be followed through the life of the individuals, since it is known positively that these effect the development and characteristics of organisms. They will be traced, as material bodies, throughout the life of the individual, and into the next generation. Understanding of the action of the genetic system is largely dependent on having a clear picture of these matters.

To present this history clearly and systematically it will be best to set forth the main facts in a series of propositions, which will be further emphasized by giving to each a number or letter. We begin at the earliest condition of the young individual; that is, at the fertilized egg before it has divided.

3

Relation of Genetic System to Characteristics

X-CHROMOSOME AS TYPE

In the preceding chapter it has been seen that differences in sex are produced by altering the chromosome combinations present in the fertilized egg, and this has brought to light some of the methods by which the chromosomes act on development and characteristics. We now Examine other effects of altering chromosomes.We shall have to examine the effects of altering all the different classes of chromosomesýÿthe X-chromosome, the Y-chromosome, and the different autosomes. Each of these three groups (X, Y, and autosomes) gives rise, as we shall see, to a different type of inheritance. The simplest relations, from the experimental point of view, are presented by the X-chromosomes. We therefore deal first with these.

EFFECTS ON CHARACTERISTICS PRODUCED BY ALTERING THE X-CHROMOSOMES

The X-chromosomes take such a course in passing from generation to generation that it is possible to follow the descendants of a particular Xchromosome (that present in the original male parent, for example), knowing in which individuals they are present, in which they are absent. Furthermore, in certain individuals (males, in organisms of Group I) there is but a single X-chromosome instead of a pair. These relations make it a relatively simple matter to discover the distinctive effects of a particular X-chromosome. The effects of X-chromosomes will therefore be dealt with somewhat fully as a type of chromosomal action. Later the action of the other chromosomes will be taken up.

METHOD OF INHERITANCE OF CHARACTERISTICS RESULTING FROM MODIFICATIONS OF X-CHROMOSOMES

The X-chromosomes, as we saw in earlier chapter pass always from the father to his daughters, not to his sons. The sons get their X-chromosomes exclusively from the mother. These peculiarities make it relatively easy to discover characteristics of individuals that depend on the kind of X-chromosomes they contain. We saw in our introductory account of the discovery of the materials of heredity certain effects of altering X-chromosomes. Sometimes, as there set forth, an X-chromosome of one of the parents is defective in such a way as to cause an abnormality or defect in all the individuals that receive this X-chromosome or its descendants. Such an abnormality was bar-eyej. Characteristics of this sort, that are manifested in every individual that has the defective X, are called dominant characteristics.

It was further set forth that there are other characteristics which result from defective X-chromosomes, but which are manifested exclusively in individuals that carry only X's that are defective; if a normal X is present also, the characteristic does not appear. Such a characteristic was white eyes. Seemingly the defective X fails to produce certain required materials. But, if a normal X is present, it produces the materials, so that no defect results. Such characteristics, that are manifested only if a normal X is not present, are called recessive.As we shall see in detail, many different characteristics are known that are thus due to defects or alterations in certain X-chromosomes. Such characteristics show very remarkable rules of inheritance. A number of such characteristics were known in man before their relation to chromosomes were discovered; their method of inheritance was extremely puzzling. It will be worth while to observe the rules of inheritance of such characters, as they appear when the relation to X-chromosomes is not brought out. In man there has long been known a defect called haemophilia, which follows this method of inheritance. It is due to a defective X-chromosome, and shows itself in the fact that the blood does not coagulate on exposure to the air, so that if wounded the defective individuals are likely to bleed to death. The defective X-chromosome fails to produce certain materials that are required for coagulation of the blood. But if a normal X is present in addition to the defective one, the necessary material is produced and the blood coagulates normally. Haemophilia is thus a recessive character. Programmes-blindness is another recessive characteristic that is due to a defective X-chromosome.

The following are the rules of inheritance shown by such a recessive character resulting from a defective X-chromosome.

1. The individual affected by the abnormality is usually a male.
2. When such an affected male mates with a normal female, none of the children are affected. It appears as if the defect were not inherited.
3. But in the grandchildren the defect reappears; it has 'skipped a generation.
4. But not all the grandchildren have the defect. There is a peculiar distribution of the defect among them, as follows:

  a. None of the sons' children have the defect; nor does it reappear in any of their descendants. So far as the descendants through the sons are concerned, the defect has disappeared.
  b. But some of the daughters' children have the defect.
  c. None of the daughters' daughters have the defect.
  d. But some of the daughters' sons have it. On the average it turns out that about one-half of the daughters' sons have the defect.
5. In later generations it continues as a rule in the way we have described, that is:
  a. Only males have the defect.
  b. It skips a generation, and reappears only in part of the grandsons through the daughters.
  c. It is thus inherited through normal females.
  d. But it is never inherited through normal males.

  The above is the usual course of inheritance; it holds for the great majority of cases.
6. But observers were perplexed by finding that in very rare cases all these rules are broken. The following were then observed:
  a. The individual affected is a male, as before.
  b. He marries a normal female; and now some of the children are affected.
  c. But not all the children are affected; on the average about one-half of all.
  d. Those affected include both males and females, in equal number.

  Thus here we find that the rule that only males are affected does not hold; females also may have the defect.
7. When now one of these rare defective females is mated with a normal male:
  a. All the sons produced are affected.
  b. None of the daughters are affected.

Thus 'the sons inherit from their mothers, the daughters from their fathers', in such a case.

One further important fact was observed:

8. If the mother and father are both affected, then all children are affected, including daughters as well as sons.

Imagine trying to get intelligible rules of inheritance from such a set of seemingly contradictory observations! Even the rules commonly followed seem arbitrary and incomprehensible. And at times they are all broken; sometimes one result is produced, sometimes another.

But when it was discovered that such defects are due to a defective X-chromosome, and that they appear wherever the defective X-chromosome has no normal one with it, all this was at once cleared up and became intelligible.

This is revealed at once when we place in the diagram the X-chromosomes of the individuals, marking in a special way the X- chromosome of the individual originally affected and in the same way all the X-chromosomes derived from that chromosome. Every individual that contains only the defective X is itself defective, whether male or female. Every individual that contains in addition a normal X is without the defect. Such defects usually appear in males only, because males have only one X-chromosome; and if that is defective, the individual is defective. Females, on the other hand, have two X-chromosomes, and as normal X's are more common than defective ones, usually any female that has a defective one has a normal one also, and is therefore not personally defective.

But rarely it happens by chance that a female gets two defective X-chromosomes; then, having no other, she is personally defective. And in that case, since she always gives one of her X's to her sons and they have no other, her sons are all defective.

Indeed, when we recognize that such a characteristic as haemophilia or programmes-blindness is due to a defect in certain X-chromosomes, but is not manifested if a normal X-chromosome is present, we discover that all the results enumerated above, and illustrated in the diagram of are inevitable.

Many other recessive characteristics, as we shall see, are due to defects or modifications of X-chromosomes. Such characters are distributed to the offspring in the way illustrated above; they all follow the same rules of inheritance. They follow the defective X wherever it goes, being manifested whenever the defective X is the only kind of X that is present. Such characters, since they are commoner in males, are commonly called sex-linked characteristics.

In the case of certain serious defects, these rules are modified by the fact that individuals with two defective X's (females) cannot live and develop. Haemophilia, for example, is not certainly known to occur in females. Any female that gets two X's defective in this way simply does not live. Programmes-blindness, however, illustrates well all the rules set forth above, and the same is true for many such defects in animals.

ABNORMAL DISTRIBUTION OF X-CHROMOSOMES

What will happen if the X-chromosomes by accident become irregularly distributed? We know that sometimes the X-chromosomes are indeed irregularly distributed. Will the sex-linked characteristics continue to follow them; will the defective characteristics show the same irregular distribution as do the X's?

Such cases in great number have been fully observed. It is found that the sex-linked characteristics do indeed follow the X-chromosomes wherever they go. This proves conclusively (if there were any possible doubt in view of the extraordinary course normally followed by such characters) that it is indeed the X-chromosomes on which the characters depend. The matter is one of importance and interest, so that it will be worth while to examine carefully certain typical cases of the result of irregular distribution of X-chromosomes.

NORMAL DISTRIBUTIONS

Sometimes the body of Drosophila is yellow instead of the normal grey; this is a recessive character due to a defect in the X-chromosomes. Suppose that we mate together a female that has a yellow body and a male that has the normal grey programmes. The female has two modified X-chromosomes, which we may represent by XX. The male has its X unmodified; we may represent its cells as XY. Normally in forming the germ cells, the two X's of the female separate into different germ cells, and all the ova receive one of the modified X's. The normal male produces germ cells of two classes, X and Y. When the germ cell X from the female unites with Y from the male, sons are produced, with the constitution XY; while X from the female with X from the father gives daughters XX. Since the yellow programmes dependent on X is recessive, the daughters XX have the normal grey programmes, like the father. But the sons XY have only one X, and as this is the defective one, from the mother, the sons are yellow, like the mother. The normal result of such a mating therefore is that all the sons are recessive like the mother, all the daughters dominant, normal, like the father. This is what occurs when the X-chromosomes are distributed to the germ cells in the usual way.

ABNORMAL DISTRIBUTION

Sometimes, as seen on previous pages, in forming germ cells both X's of the female go together to a single ovum, while other ova receive no X. In a race discovered by L. V. Morgan the two X-chromosomes in the female were partly united, so that they almost invariably thus go together to one ovum, leaving other ova without an X. In this race the X-chromosomes were so modified as to produce the recessive yellow programmes described in the preceding paragraph. We therefore have an opportunity to determine the course of inheritance of the yellow and the grey programmes when the X-chromosomes are thus abnormally distributed.

Represent as before the X-chromosomes that produce the yellow programmes by X. Thus one set of ova receive XX, while another set receive no X. When the germ cells XX united with the germ cells Y from the normal father, there were produced daughters XXY, in which both X's came from the mother, and these daughters were yellow like the mother (since no normal X was present). Sons were produced by the union of ova that contained no X with sperms that carried X; this gave sons XO, with their single X from the normal father, instead of from the mother, as is usually the case. And such sons had the normal grey body programmes of the father, instead of the yellow body programmes of the mother, as happens normally. Thus with a change in the distribution of the X's there is a corresponding change in the method of inheritance. If the sons receive their X from the recessive mother, as in the normal cases, they are recessive like the mother. But when, through non-disjunction, the sons receive a dominant X from the father, they are dominant like the father. Similarly, in the normal case the daughters receive a dominant X from the father and are therefore dominant like the father. But in cases of non-disjunction the daughters receive both their JPs from the recessive mother, and are therefore recessive like the mother. In sum, when the X's are normally distributed, such matings give 'criss-cross inheritance'; sons like the mother, daughters like the father. But when the X's are not so distributed, there is no criss-cross inheritance; sons are like the father, daughters like the mother. The dominant and recessive characters follow the respective X-chromosomes, whether these are distributed normally or abnormally.Such experiments have been repeated many times, and with other sex-linked characters. Always the characteristics follow the distribution of the X-chromosomes, however these are distributed.It may be concluded with certainty that it is the presence of X-chromosomes of a modified type that causes the appearance of the particular sex-linked characters that are manifested. The rules of distribution of sex-linked characters are the rules of distribution of the X-chromosomes.

DOMINANCE AND RECESSIVENESS IN SEX-LINKED CHARACTERS

As we have seen, and shall see further, many defective conditions are due to defects in certain X-chromosomes. These follow from generation to generation the distribution of these X-chromosomes. Most of these bodily defects are manifested only in individuals in which the defective X-chromosomes are the only kind present; that is, these defects are recessive.

But it is important to observe that in such cases the normal condition of the organism likewise follows the distribution of certain X-chromosomes. Haemophilia follows certain X- chromosomes. But in the same matings in which this occurs, the healthy condition of the blood follows certain other X-chromosomes. If a defective mother is mated with a normal father, the sons are defective because they receive only the mother's defective X-chromosomes. But in the same way the daughters are normal because they receive the father's normal X-chromosomes. In the normal individuals the X-chromosome plays a part, just as it does in the defective individuals. In the normal individuals it is healthy and supplies what is required for normal development, while in the defective individuals it fails to supply what is required.

In the case of most defects this normal condition is dominant over the recessive condition; that is, when both are present, the normal chromosome is the one that prevails in its effect on the individual. There are defective characters, however, in which the defective condition is dominant or partly dominant. In these cases, when a defective X and a normal one are present together, the defective one produces its effect, as in the case of bar-eye, described earlier. In many such cases the defect is less marked when a normal chromosome is present as well as a defective one; then the defective condition is said to be partly dominant.Thus whenever an individual bearing defective X-chromosomes is mated with a normal individual, we have both dominant and recessive characters, the course of which may be followed in the later generations the dominant character being commonly the normal or usual condition. How this works out is shown in our next paragraphs.

TESTS FOR SEX-LINKED INHERITANCE

By observing the distribution of the dominant and recessive characters among the offspring of certain matings, a test for determining whether given characters depend on the X-chromosomes is supplied.

The test consists in making what are called reciprocal crosses.

1. On the one hand mate a dominant female with a recessive male;
2. Also mate a recessive female with a dominant male.

Represent a dominant X-chromosome by a capital X, a recessive X-chromosome by a lower-case letter x. In the group of organisms with which we have been dealing, the female has two X-chromosomes, the male but one. The daughters receive an X from each parent, the sons an X from the mother only.

A dominant female will be represented by XX, a recessive male by xo; similarly a recessive female is xx, a dominant male is XO. The two ma tings and their results will then be represented as follows:

1. XX by xo gives X + XO Dom. Mother Rec. Father Dom. Daughters Dom. Sons
2. xx by XO gives X + xo Rec. Mother Dom. Father Dom. Daughters Rec. Sons

Thus when the mother is dominant, the father recessive, all the children are dominant, like the mother. When the father is dominant, the mother recessive, the daughters are dominant like the father, the sons recessive like the mother; this is called 'criss-cross inheritance'.

Whenever reciprocal crosses give these results, it is certain that the two diverse characteristics (dominant and recessive) result from differences in the X-chromosomes of the two parents. 'Criss-cross inheritance' is particularly useful in showing at once that we are dealing with sex-linked inheritance. In every case where such results are produced, further tests show that the two characteristics follow in later generations the two different kinds of X-chromosomes wherever they go. It is mainly by the use of these tests that the many different characteristics dependent on diversities in X-chromosomes have been discovered.

SEX-LINKED INHERITANCE IN GROUP II.

Sex-linked inheritance as we have just described it was originally discovered in animals belonging to Group I, in which the females have two X-chromosomes, the males but one (with or without a Y- chromosome).

But when the tests we have just described were applied to birds and to certain other animals, it was discovered that there is a second group (Group II) in which it is the male that has two X-chromosomes, the female but one. In the common fowl, reciprocal crosses were made between 'barred' fowls (Plymouth Rock) and black fowls (Langshan). Here the 'barred condition was found to be dominant. The results were:

1. Barred father by black mother=Barred sons + barred daughters.
2. Black father by barred mother = Barred sons + black daughters.

Here the second mating gives 'criss-cross inheritance', showing that we are dealing with sex-linked characters dependent on differences between the X-chromosomes of the parents. But, in the first mating above, all the offspring are like the father instead of like the mother, which latter was the case in the matings on the previous pages. The father in this case has the dominant characteristic (barred), and all the offspring are dominant like the father instead of like the mother. When the mother is dominant ('barred') the result in the fowls is criss-cross inheritance; while, in the cases before described, criss-cross inheritance occurs when it is the father that is dominant. In fact, in the birds, the males and females simply exchange roles, as compared with their roles in the organisms of Group I. This appears clearly from the following comparison of results in the two cases:

Parents Offspring

Group I. Dom. Mother, Rec. Father

Group II. Dom. Father, Rec. Mother) A" offspring dommant Chromosomes ýÿ XX xo Xx ýÿ XO

1. Group I. Dom. Father, Rec. Mother:ýÿDom. Daughters, Rec. Sons
   Group II. Dom. Mother, Rec. Father:ýÿ Dom. Sons, Rec. Daughters
   Chromosomes ýÿ XO xx X* ýÿ xo

Since the roles of males and females are interchanged in the two groups, their chromosomal conditions must also be interchanged; that is, since in Group I the female has two X's, and in Group II the male plays the same role as the female of Group I, the male of Group II must have two X's, the female but one. The chromosomal conditions in the two groups are shown in the above tabulation, X signifying dominant, while x signifies recessive, and o signifies the lack of X (whether Y is present or not). Careful examination shows that this is the only way in which the results in Group II can be produced.It was in the way illustrated above that the existence of Group II was discovered. Birds and certain moths show inheritance of the kind typical for Group II. In both these it has since been found under the microscope that the males have indeed one more chromosome than the females.

THE ROLE OF THE X-CHROMOSOMES IN DEVELOPMENT

A great many characteristics, of various kinds, have been found to be the result of modifications of particular X-chromosomes, and thus to follow in later generations the distribution of the descendants of those chromosomes. Examination of a number of these is desirable, both for their own importance and for the light they throw on the functions of the X-chromosomes.In man, the following characteristics, among others, dependent on diverse types of X-chromosomes, are known from the fact that they show typical sex-linked inheritance.

Haemophilia.Lack of coagulability in the blood. This results from a serious defect in certain X-chromosomes. The existence of defective X-chromosomes having this result shows that the normal X-chromosomes play a part in supplying something necessary for producing normal blood that coagulates properly.

Programmes Blindness. The fact that defectiveness in X-chromosomes causes programmes blindness shows that the normal X's play a role in producing the normally functioning eyes.

NIGHT BLINDNESS.

Inability to see in a poor light.

Near-Sightedness of certain types.

Progressive atrophy of the muscles (Gower's disease).The fact that defects in X-chromosomes have this effect shows that the normal X's play a role in the normal functioning of the nerves and muscles.

A considerable number of other sex-linked characters are known in man. But man is a very unfavourable organism for the study of inheritance. Yet even the little that is known of sex-linked inheritance in man shows that the X-chromosome plays a role in many diverse bodily functions. For a more complete idea of its role, some organism that can be bred experimentally must be examined. For this purpose the fruit-fly, Drosophila melanogaster, is the best organism to select, since it has been studied more extensively than any other.

In the fruit-fly we find peculiarities of the following types that are dependent on alterations in particular X-chromosomes:

Many different eye colours. The normal eye programmes in this organism is a certain shade of red. Defects in different X- chromosomes result in producing, in the individuals that bear these defective chromosomes, many different shades of red, varying from deep red to a very light red, and thence to 'buff', 'ivory, and 'white'. More than a dozen different types of eye programmes are known to result from modifications of the X-chromosome. It is clear that the normal X-chromosome plays an important role in producing the normal eye programmes. Structural peculiarities of the eye, such as 'bar-eye', 'facet eye', 'furrowed eye', and the like, are known to result from modifications of the normal X-chromosome. Wing modifications. Many different conditions of the wings are known that depend upon diversities among X- chromosomes borne by different individuals. These affect all sorts of features of the wings: size, form, venation, function. The normal X-chromosome obviously plays an important role in the full and normal development of the wings.

Body colours and markings.ýÿModifications of the X- chromosomes produce the body colours yellow, sable, tan, chrome, lemon, green, and the like, in place of the normal grey. Other changes in X alter the distribution of pigment on the body, giving the characters *dot\ and the like.

Body structure.A defect in X produces irregularities in the abdomen, known as 'abnormal abdomen'.

Legs.A defect in X-chromosomes causes abnormal development of the legs; some of them are wholly or partly reduplicated, so that the total number of legs is increased.

Bristles.Modifications in X-chromosomes result in various different changes in the bristles that are scattered over the body.

Many physiological conditions and functions likewise depend upon the X-chromosomes, since they are changed when these chromosomes are modified. Among these are the following:

Positive reaction to light. Drosophila with normal X-chromosomes fly Towards a source of light. Those having defective X-chromosomes of the kind that produce a tan-coloured bodydo not fly Towards a source of light. Weakness and short life. Most of the, defective X-chromosomes that cause structural or other changes in the body (different eye colours, wing forms, and the like) produce likewise weakness and short life. The individuals bearing them are less resistant to bad conditions, and live for a shorter time than the individuals that bear the normal X-chromosomes.

Life and death. Some X-chromosomes have defects that are so severe that the individuals bearing them will not live and develop, unless there is present also a normal X-chromosome. Such defects are known as lethals. The presence of such lethal defects in the X-chromosome has the result that males bearing them die, since they have but one X; while females usually live, since they have an additional X that is often normal. If a mother that has one lethal X and one normal one is mated to a father that has a normal X, the result is that half of the sons which receive the mother's defective X fail to develop, while the other half of the sons,receiving the mother's normal X, live and develop. All the daughters live, since they all receive a normal X-chromosome from the father. The consequence is that in such families there are twice as many daughters as there are sons. In families of 100 to 200, as occur in Drosophila, this is very striking.X-chromosomes are known to play in other organisms roles similar to those mentioned above for Drosophila, although in no other organism has the matter been so fully studied.

We may summarize what has been brought out above in the following statements:

1. Many defects that appear in individuals are the result of defects in the X-chromosomes that they bear.
2. Some of these defects are dominant; they are manifested in all individuals in which the defective X is present.
3. But most such defects are recessive; they are manifested only in individuals in which the defective X is the only kind present.
4. When both a normal X and a defective X are present, in most cases the normal X performs the required functions, so that the individual is not defective.
5. Since the female has two X-chromosomes, while the male has but one, such defects are more frequently manifested in the males. Usually one of the X's present in the female is without the defect, so that she is not defective. But when the male carries a defective X-chromosome, the defect is manifested.
6. It is thus advantageous, so far as the occurrence of defects is concerned, to have two X-chromosomes rather than one.
7. This gives the female a considerable advantage over the male in these respects. There are many different types of defects due to defective X-chromosomes; almost all of them are more common in males than in females. Some of these defects seriously injure the health, or even result in death, if the defective X is the only kind that is present. It is well known that, in general, the death rate is higher in males than in females; this is probably due to the fact that males have but one X-chromosome.
8. Among the different individuals of a species are scattered a great number of different types of X-chromosomes having different effects on development. Many of the diverse X-chromosomes are distinctly defective, causing personal defects in the individuals that bear them. Others are not defective.
9. Defects or modifications in X-chromosomes affect in different cases all parts and functions of the organism.

THE ROLE OF THE X-CHROMOSOMES IN DEVELOPMENT

What do the facts brought out in the preceding sections show as to the physiological functions of the X-chromosomes; as to their role in development?

For every kind of effect resulting from defects or modificacations in the X-chromosomes, there is a corresponding (opposed or diverse) action of normal or unmodified X- chromosomes. Since defective X-chromosomes of a certain type produce in Drosophila short, ill-developed wings, it follows that the normal X-chromosomes act on the development in such a way as to give long, well-developed wings.For if we substitute the normal X for the defective one, this causes the normal well-developed wings to be produced. Similar reasoning applies to all the defective conditions that result from defective X's; the normal X's act in such a way as to produce the corresponding normal conditions.

It follows therefore that the normal X-chromosomes play a role in the production of eye programmes and structure, body programmes and structure, in the development of the legs, the wings, the bristles, in producing normal health, resistance and vigour, and in various physiological processes. We know, further, that they play a most important part in determining sex, with all that this includes of structural and physiological influence.

Clearly therefore the X-chromosomes play a most important part in development. Their action is not limited to any single part of the body, nor to any single class of functions. They enter into the processes of development in such a way as to influence all parts of the body, all functions of the body. They begin to influence development very early, as was seen in considering the development of sex. And they play important roles in such relatively late processes as the production of eye colours. It can hardly be doubted that they enter into the developmental activities from practically the beginning, influencing all the later processes that occur.

AUTOSOMES AND T-CHROMOSOMES

In the two preceding chapters the X-chromosomes have been dealt with because of the great advantages they offer for experimental study. It was seen that they affect development in many ways, and that in a given species there are many different types of X-chromosomes in the different individuals, causing them to show different characteristics, structural and physiological.

Is the X-chromosome typical in these relations? Shall we find similar effects in the other chromosomes?

In addition to the X-chromosomes, there are the autosomes, which are present as a rule as several or many pairs. In man there are 23 pairs of autosomes: in Drosophila there are three pairs. There is, further, in many species the single Y-chromosome. The autosomes and the Y-chromosome are distributed from parent to offspring in characteristic ways, differing from the distribution of X- chromosomes. If they affect characteristics, these differences should appear in a different system of inheritance. We shall examine first the autosomes.

THE AUTOSOMES: TYPICAL MENDELIAN INHERITANCE

We have already seen that the autosomes play a role in development, in the fact that they help to determine the sex of the developing individual. In Drosophila, if the number of sets of autosomes, in relation to the number of X's, is changed, this changes the sex.Do autosomes also affect other characteristics? Do the different pairs of autosomes have different functions? Are there diverse types of autosomes, with different effects, in different individuals of a species, as is the case in relation to the X-chromosomes? Are there dominant and recessive effects of autosomes, as there are of X?

If dominant and recessive characteristics depend on autosomes, then these characteristics must follow the distribution of the autosomes from parents to offspring, as the characteristics dependent on X follow its distribution. We must then examine the method of distribution of the autosomes, to see whether there are characteristics that show this method.

METHOD OF DISTRIBUTION OF THE AUTOSOMES

The distribution of the autosomes from parents to offspring is much simpler than that of the X-chromosomes. The main facts are as follows:

1. The autosomes are in pairs in all individuals. There are no individuals with unpaired autosomes, like the male of some organisms with relation to X.
2. Thus the two sexes are alike as to their autosomes.
3. In forming germ cells, one member of each pair of autosomes goes to each germ cell. The sperms and the ova are alike with respect to the autosomes.

   Thus the fundamental rule with relation to the autosomes is that each germ cell carries one member of each pair of autosomes.
4. By union of two germ cells the pairs of autosomes are restored.
5. Thus each one of the offspring carries in each pair of its autosomes one autosome from each parent.

Now consider a single pair of autosomes, such as the pair marked AA in. Suppose that in different individuals the members of this pair of autosomes produce different effects, as we have seen to be the case with the X-chromosomes of different individuals. And suppose that, as in the case of X, one of these effects is dominant, the other recessive. How would such characteristics be inherited?

This question must be examined with care. For this purpose we may take as an example the different colours found in different varieties of peas, used in the classic experiments of Mendel. The yellow seed programmes of a certain variety is dominant, and results from the constitution of the autosomes of a certain pair. The green seed programmes of another variety is recessive, owing to a different constitution of the autosomes of that same pair. Call the two autosomes of that pair in the dominant yellow race by the capital letters AA, while the two autosomes that give rise to the recessive green programmes in the other race may be called aa. The dominant parent may be designated D, the recessive parent R, When these two parents produce germ cells, each germ cell receives, of course, one chromosome for the pair. The germ cells from the dominant parent D have the autosome A, those from the recessive parent R have the autosome a.The germ cells A, from parent D, unite each with a germ cell a from parent R giving offspring (zygotes) containing both A and a. These first generation offspring are commonly designated Fi (first filial generation).

And since the characteristic carried by the autosome A is dominant, all these Fi offspring manifest the dominant characteristic D (yellow, in case yellow-seeded peas are crossed with green-seeded peas).

Next observe the result of mating together two of these individuals Aa, of the Fi generation, giving the Fa (second filial generation).

Each parent Aa gives, according to the general rule, two kinds of germ cells, one kind carrying the dominant autosome A, the other the recessive autosome a. Each kind of germ cell from one parent unites with each kind from the other. That is, half the A germ cells from one parent unite with A germ cells from the other parent, half with a germ cells from the other parent; and the same is true for the a germ cells from the first parent. This gives in equal proportions Fa offspring (zygotes) as follows:

AA + Aa + a A + aa

But as the two combinations Aa and aA contain the same chromosomes, the proportions may be written AA + 2Aa-f-aa

The individuals of the constitution AA and Aa contain the dominant chromosome A; therefore they manifest the dominant characteristic, D. The individuals aa, having only the recessive chromosome a, manifest the recessive characteristic, R. Thus in this F2 generation, so far as manifested characters are concerned, the proportion of offspring are 3 Dominant to i Recessive

Thus if dominant and recessive characteristics depend on the two members of a pair of autosomes, when a dominant individual AA is mated with a recessive individual aa the general result is bound to be that the immediate offspring (Fi) all manifest the dominant characteristic, while in the Fa generation (produced by mating two individuals from Fi) there will be three times as many dominant individuals as there are recessives.

Now, this is the result that was discovered by Mendel in 1866; it formed the foundation for the working out of Mendelian heredity. The proportion 30 to iR in the Fa generation is commonly known as 'the Mendelian ratio'; it is one of the most characteristic proportions in this type of inheritance.

Mendel discovered that there are many different characteristics that yield this 3 to i ratio. If yellow and green peas are crossed, in the Fa generation there are 3 yellows to i green; if round peas and wrinkled peas are crossed, in F2 there are 3 round to i wrinkled; and so for many other characteristics in the pea plants with which Mendel worked.

And since the time of Mendel these same ratios have been found for thousands of different characteristics, in great numbers of different organisms, from the pea plant and Droso- phila to man. The fact that inheritance gives these ratios is commonly spoken of as Mendel's law, and the type of inheritance which it exemplifies is spoken of as Mendelian inheritance. The type that gives in the Fa generation 3D to iR is the commonest method of inheritance, much commoner than the sex-linked method, in which the characteristics depend on differences in the X-chromosomes. This is obviously because autosomes are more numerous than X-chromosomes.We find then that inheritance according to Mendel's law takes place exactly as it would if the characteristics concerned depended on the two members of a pair of autosomes. As will be shown later, it can be proved experimentally that such characteristics do indeed depend upon autosomes.

Thus the rules of Mendelism are the rules of distribution of the two members of a pair of autosomes, one having a recessive effect, the other a dominant effect. Mendelian heredity is heredity of characteristics that depend on differences in autosomes.Mendel did not know of the relation of the characteristics to the chromosomes. He discovered, however, certain other important facts about this type of heredity, facts which have become clearly explicable since it has been learned that this type depends on differences in autosomes. These facts must be examined.

For this purpose it will be convenient first to define certain terms which are universally employed in dealing with inheritance.

By the union of two germ cells or gametes, as A and a, there is produced a zygote. If the two germ cells are alike with respect to the chromosomes or characteristics that they carry, the zygote so produced is spoken of as a homozygote; thus AA and aa are homozygotes. If the two germ cells that unite to form a zygote are unlike, as A and a, the zygote so produced is called a heterozygote, as Aa.

The two corresponding characters, one dominant, the other recessive, that are produced by the two differing members of a pair of chromosomes are spoken of as alternative characters, or alleles, or allelomorphs. Any given individual manifests one or the other of the two alternatives, not both. For the alternative characters we shall employ henceforth the term alleles.

The characteristic manifested by a zygote is its phaenotype^ while its actual constitution is its genotype. Thus, two zygotes AA and Aa both manifest the dominant character; the dominant character is the phaenotype, or the individuals are said to be phaenotypically dominant. But they are diverse in genotype; one is AA, the other Aa. In the case of a homo- zygote AA or aa, the phaenotype and genotype may be held to coincide.

Mendel crossed yellow peas with green peas, and found, as before stated, that in the Fi generation all the peas were yellow, so that yellow is the dominant allelomorph. In the F2 there were three times as many yellows as greens. The actual number of seeds in the F2 generation was in his experiments 8023, of which 6022 were yellow while 2001 were green, so that the ratio of dominants to recessives was, accurately expressed, 3.01 to i. Mendel found that if he bred further the individuals of the F2 generation, some of these gave different results from others. In the pea plant it is possible to allow each plant to fertilize itself, so that both kinds of germ cells come from the same parent. When this was done with the individuals of the F2 generation, the following results were observed:

The green plants (recessive) gave only recessive green offspring (they 'bred true').

Of the yellow plants (dominant), one-third gave only yellow offspring (they 'bred true'), while two-thirds of them gave both yellow and green offspring, in the proportion of 3 yellows to i green.

The reason for this is clear when we consider the chromosomes on which the characters depend. If A represents the chromosome (autosome) that gives the dominant yellow programmes, while a represents the chromosome that gives the recessive green, the original parents were AA and aa, the Fi generation were all Aa, while the Fa generation showed the proportions AA -f 2 Aa + aa. The recessive individuals aa are homozygotes; their two chromosomes are alike, being aa. When germ cells are produced by these homozygotic individuals, all these germ cells must contain the chromosome a, and when two of these unite in self-fertilization, of course they produce in all cases again individuals aa. These are, of course, recessives (green), like their parents; that is, the green parents aa 'breed true'.

Similarly the dominant yellows, AA (constituting one- third of all the dominants), are homozygotes. Their germ cells all contain the chromosome A, and the offspring produced by union of such germ cells, in self-fertilization, are all AA, and so dominant yellow like the parents. One-third of all the dominants thus breed true.

But the individuals Aa are heterozygotic; that is, they have two kinds of chromosomes, A and a, in the pair. Thus they produce two kinds of germ cells, A and a, in equal numbers. And when these germ cells unite, they necessarily give, as we saw on an earlier page, offspring of three kinds, in the proportions AA + 2 Aa 4- aa Of these, AA and 2Aa will be dominants (yellow), while aa will be recessive (green).

Thus the homozygotes aa and AA breed true, while the heterozygotes, Aa, do not.

Furthermore, as Mendel discovered, if the heterozygotes Aa are mated with the recessive homozygotes aa, they give two kinds of offspring, Aa and aa; half (Aa) being dominant,

the other half (aa) being recessive. If the heterozygotes Aa are mated with the dominant homozygotes AA, the offspring AA and Aa are all dominant, though half (AA) are homozygotes, the other half (Aa) heterozygotes.

All these results are typical for the descendants of two parents different in a single pair of chromosomes, one pair being dominant, the other recessive. For reference it will be well to make a table of these results, characteristic for Mendelian inheritance:

THE RESULTS OF MENDELIAN INHERITANCE

Parents: AA and aa; both homozygotic. F i: Aa; all alike, dominant heterozygotes.

Fs: AA + 2 Aa + aa. Three-fourths of the individuals dominant, onefourth recessive. Half the individuals homozygotes (AA and aa), the other half heterozygotes (aAa).

Each type of homozygote bred by itself produces offspring like itself.

The heterozygotes Aa, when bred by themselves, produce again offspring

AA + 2 Aa + aa Aa mated with aa yields again Aa + aa Aa mated with AA yields again Aa + AA. All of these results are found to hold true for thousands of different characteristics.

We may therefore summarize the situation in the following general propositions:

1. When two different individuals are crossed, in many cases the differences between them follow the same method of distribution as do the two members of a pair of autosomes ýÿone being dominant, the other recessive.
2. This distribution is what is known as typical Mendelian heredity.
3. The two sexes do not differ in these respects. Each sex has the same proportion of dominants and recessives.
4. The rules of Mendelian inheritance are essentially the rules of distribution of the members of pairs of autosomes.

Characteristics that depend on two or more pairs of autosomes.We have thus far followed the distribution of characteristics that depend on differences in a single pair of autosomes. But most organisms have several or many pairs of autosomes: the fruit-fly has three pairs, man has twenty- three. Do some characteristics depend on differences in one pair of autosomes, others on differences in other pairs? Have the different pairs of autosomes thus different functions in development and inheritance?

We find that the different pairs of autosomes have indeed different functions, and that some characteristics depend on one pair, some on another. This is shown in many cases in which two parents differ in two diverse characteristics. One characteristic may then act as if dependent on one pair of autosomes, the other on another pair. As an example of this, we may take a case studied by Mendel in peas.One of the original pea plants had seeds that were (i) yellow, and (2) round, or smooth. These, as it turns out, are both dominant characters. The other plants, crossed with those just mentioned, had recessive characters; the seeds were (i) green, and (2) wrinkled.When these two kinds of pea plants were mated together, it turned out, as we shall see, that yellow and green behave in inheritances as if due to a difference in the two parents in the effects of a certain pair of autosomes; that is, they are alleles. Round and wrinkled behave as if due to a difference between the two parents in the effects of another pair of autosomes. We may therefore represent the two pairs of autosomes. Representing the dominant characters by capitals, the recessives by lower-case letters, we may in the diagram represent the chromosomes giving origin to the four characters as follows:

A=yellow a=green

B = round b=wrinkled

The parents will then be represented as shown in Fig. The germ cells from these parents contain, of course, one member of each pair, as shown in Fig. One parent produces gametes all of the constitution AB, the other gametes all of the constitution ab. When a gamete AB from one parent unites with a gamete ab from the other, all the offspring (Fi) thus produced have the constitution AaBa, as shown at Fi in Fig.. Since the dominant autosomes A and B are both present, all these individuals are dominant for both pairs of characters; that is, they all have seeds that are yellow and round, as Mendel found. But all are heterozygotes-for both of these characters.

Next we mate together two of these Fi individuals to give the F2 generation. In forming gametes the rule is, of course, that each gamete contains one member of each pair of chromosomes. But the two pairs, Aa and Bb, are distributed to the gametes quite independently.

Thus, some get A and B, some A and b, some a and B, some a and b. There are thus from each parent four types of gametes, AB, Ab, aB, and ab, each type being represented by many gametes. The gametes of each of the four types from one parent unite in approximately equal proportions with gametes of each of the four types from the other parent. That is, gametes of the type AB unite in different cases with AB, Ab, aB or ab from the other parent; and so of each of the other gametes of the first parent. The result is to give a set of 16 combinations of gametes (some of which are alike, however, as will be seen). It is not necessary to carry further in our diagram the outlines of the chromosomes.

INHERITANCE DEPENDING ON TWO PAIRS OF AUTOSOMES

AB AB AB AB

AB(i) Ab(6) aB(7) ab(s)

Ab Ab Ab Ab

AB(6) Ab(2) aB(5) ab(8)

aB aB aB aB

AB($_7$) Ab($_5$) aB($_3$) ab($_9$)

ab ab ab ab

AB($_5$) Ab(8) aB($_9$) ab($_4$)

But, as will be observed, not all these 16 are diverse; thus ABAb is the same in constitution or genotype as AbAB. If we number the different combinations, and give the same number to those that are alike, as is done (in parentheses) in the table, we find that there are nine different combinations or genotypes. That is, two parents that are both ABab produce offspring of nine different constitutions or genotypes, in the relative proportions given in the table. When the parents differed in but one pair of characters there were but 3 different combinations. When there are two pairs the number is 3>or9-

These nine different combinations will manifest the dominant characters A or B, in all cases where an A or B is present whether present once, as A, or twice, as AA); they will manifest the recessive characters a or b in all cases in which a or b is present twice, no corresponding capital letter being present. Collecting together these different types, we find the different characteristics to be manifested in the following proportions of the zygotes:

9 AB + 3 Ab + saB +1 ab

That is, 9 out of 16 are dominant for both pairs of characters, 3 are dominant for A, recessive for b; 3 dominant for B, recessive for a, and i out of 16 is recessive for both.

There are thus 4 different phaenotypes. In Mendel's experiment with peas, in which the two pairs of characters were (1) yellow and green; (2) round and wrinkled; in Fi the proportions were 9 that were yellow and round, 3 that were yellow and wrinkled, 3 that were green and round, and i that was green and wrinkled. These proportions hold generally when there are two pairs of characters connected with different pairs of autosomes.

It may be observed that the proportions of the individuals of different constitution or genotype in F2, descended from a cross between parents that differ in respect to two pairs of autosomal characters (so that the parents are AABB and aabb), can be found directly from the proportions for a single pair of characters. In Fa, for the single pairs the proportions are AA + 2Aa-haA and BB + 2Bb + bb respectively.

If now we multiply together algebraically these two polynomials, we obtain the relative proportions for the different combinations produced when the two pairs are together. These proportions are the following:

AABB + 2 AaBB + aaBB + 2 AABb + 4AaBb +

(1) (7) (3) (6) (5)

2aaBb + AAbb + 2 Aabb + aabb

(9) (2) (8) (4),

To bring this out, the same numbers employed in table 2 are placed in parenthesis beneath each combination. Certain other relations among the nine different types are important, since they hold for all cases of two pairs of characters that are dependent upon different pairs of autosomes. These are the following:

1. There are four combinations (Nos. 1, 2, 3, 4), constituting one-fourth of all the zygotes, that are homozygotes for both pairs of genes. The four homozygotes are diverse in their combinations and in the characters they manifest. They form in table 2 the diagonal from upper left to lower right. Any one of these when self-fertilized or mated with another like itself breeds true: that is, it produces offspring that have the same constitution as the parents.
2. Four out of the 16, or one-fourth of all the zygotes, are heterozygotes for both pairs of characters. They form the class numbered (5), and constitute in table 2 the diagonal from upper right to lower left. Any of these when self-fertilized or mated with another like itself gives again the nine different types, in the proportions above set forth.
3. The one-half of all the zygotes, are homozygotes for one of the two pairs, heterozygotes for the other. Four are homozygotic for A and heterozygotic for B; the other four are heterozygotic for A and homozygotic for B. In these 8 there are but 4 diverse combinations or genotypes.

If any one of these is self-fertilized, it yields 3 diverse types with respect to the pair for which it is heterozygotic. Thus, the class (6), having the constitution AABb, yields when self- fertilized three combinations in the following proportions:

AABB + 2AABb + AAbb

The above relations hold for all cases in which two pairs of characteristics are dependent on differences in two different pairs of autosomes, whatever the method of designating the characters or chromosomes.In a similar way, two parents may differ in three characters that are dependent on three different pairs of autosomes. Such parents could be represented as AABBCC and aabbcc.

Their progeny (Fi generation) would all have the constitution AaBbCc. Such Fi individuals of course produce a variety of germ cells, in accordance with the principles that each germ cell receives one member of each pair of autosomes, and that the different pairs of autosomes are distributed independently. Each parent AaBbCc thus yields the following 8 types of germ cells: ABC, ABc, AbC, aBC, Abe, aBc, abC, abc.When each of the 8 kinds from one parent unites with each of the 8 kinds from the other, there are produced, of course, 64 groups, which can be arranged in a table like the 16 groups of table 2. If we classify the 64 individuals by the characters which they manifest, we find that there are 8 different phae- notypes manifesting respectively the following characteristics:

ABC, ABc, AbC, aBC, Abc, aBc, abC, and abc

That is, some are dominant for all three characters; some for a particular two and recessive for the other; some are dominant for a particular one and recessive for the other two; and some are recessive for all three. The different phaeno- types are present in different proportions; the proportions when a large number are examined are as follows:

2?ABC + gABc + gAbC + gaBC + $Abc + saBc + 3abC+iabc

The total number of different combinations (genotypes) that occur in such a case is 27, or 3. These combinations can be obtained in their correct proportions by multiplying together algebraically the expressions:

Similar but still more complex series are obtained if parents are interbred that differ in four or more characters depending on four or more different pairs of autosomes. If the number of pairs of autosomes is called «, the following relations hold:

If parents differ in n autosomes:

1. The number of diverse types of gametes produced by the individuals of the Fi generation is 2^n.
2. The number of different combinations (genotypes) in the zygotes of F2, formed by mating the 2^n gametes from Fi,iss^{w}.
3. The number of different phaenotypes produced in Fa is2^n.

These relations yield very great numbers of diverse combinations, in cases where the number of pairs of autosomes is large.

Thus in man, with 23 pairs of autosomes, if all the autosomes differed in the two parents, the number of types of germ cells produced by the Fi individuals would be 2, or many millions, the number of possible diverse genotypes in F2 would be 3, or many billions, and the number of possible different phaenotypes would be 2. (As will appear later, the number of different possible combinations is greatly increased by the exchange of parts by the members of a pair of chromosomes.)

Thus typical Mendelian inheritance, dependent on differences between different pairs of autosomes, can be treated to a certain extent mathematically. The individuals in F2 and later generations occur in proportions represented by simple mathematical ratios.

It will be worth while to formulate the main facts that we have brought out as to autosomal inheritance:

1. Some single pairs of characters, dominant and recessive, follow in inheritance the method of distribution of a single pair of autosomes.
2. Some pairs of inherited characters follow the rules of distribution of two or more different pairs of autosomes, in the following particulars:
   a. One member of each pair goes into each germ cell.
   b. The members of the different pairs are distributed to the germ cells independently, so that all possible combinations containing one member of each pair are formed with equal frequency.
   c. Each type of gamete from one parent unites with each type from the other parent, the different matings being equally frequent.

3. These methods of distribution result, for the characters manifested in the Fs generation, in certain typical numerical proportions or ratios, known as Mendelian ratios.
4. This method of inheritance is known as Mendelism, or Mendelian inheritance. The rules of Mendelism are the result of the rules of distribution of pairs of autosomes.
5. Most characteristics of organisms are inherited in this way; many more than are inherited in the sex-linked manner. This is evidently because there is but a single pair of X-chromosomes, while there are several or many pairs of autosomes.
6. Thus it is clear that, among the different individuals of a species, there are many different types of autosomes (as there are different types of X-chromosomes), producing many kinds of characteristics.

THE T-CHROMOSOME

There remains another chromosome, not included in the autosomes: namely, the Y-chromosome. Has this functions similar to those of X and of the autosomes? Does it too play a role in heredity?

As we shall see, the Y-chromosome differs in many ways from the other chromosomes. Let us first summarize the main facts as to the occurrence and distribution of Y-chromosomes.

1. Some organisms have no Y-chromosomes. Among those are many insects the dog, the cat, the horse.
2. In some organisms it is small and appears to be degenerate. This is the case in many insects; also in man.
3. In some organisms the Y-chromosome is large, but of a different form from its mate X. Such is the case in Drosophila
4. In some organisms Y is not distinguishable from X inform or size, but experiment shows that it differs from X in function.
5. The Y-chromosome descends from father to son only, exclusively in the male line. Normally it never occurs in females.
6. Hence any characters that depend on the presence of a particular kind of Y will *(a)* be found in males only, and will *(b)* never be inherited *through* females (since Y does not pass through females).

THE T-CHROMOSOME IN DROSOPHILA.

The functions of the Y- chromosome have been more thoroughly studied in Drosophila than in any other organism.

Most of the facts as to this have already been brought out, in connection with our account of the functions of the X-chromosome. We may summarize as follows the known facts as to the functions of the Y-chromosome in Drosophila:

1. The Y-chromosome, although it normally occurs in males, is not required for the production of male individuals. Through non-disjunction, male individuals are produced that contain no Y.
2. But the presence of Y is necessary for the fertility of males.
3. Y may exist in females, as a result of accidental nondisjunction; it does not affect their sex.
4. During many years of intensive study no characteristics were found that follow the Y-chromosome.
5. But recently it has been shown that the presence of Y prevents the appearance of a certain recessive character ('bobbed') that is due to a modified X-chromosome. This modified X results in the production of shortened bristles on the body of the fly. But this peculiarity does not occur if a Y is present, so that Y has the dominant normal allele for 'bobbed'. Indications of certain other ill-defined functions of the Y-chromosome have of late come to view.
6. No diverse modifications or defects such as are known in great number for the X-chromosome are known in the Y-chromosome.
7. Thus in Drosophila the Y-chromosome appears to have little function; it seems nearly but not quite inert.

Active T-chromosomes in certain Organisms; Relation to Male Secondary Sex Characters.In certain organisms the Y-chromosome is more active and varied in its effects. In such cases it shows in different individuals diverse modifications, comparable to the diverse types of the X-chromosome in Drosophila, and these produce diverse inherited characteristics that appear in males only.

This matter is closely bound up with the question of what produces the diversity of secondary sex characters. In many animals there are special characteristics which, like the Y- chromosome, are found only in males. These are the male secondary sex characters: horns, mane, beard, greater size, diversities in form, and the like. Since these appear only in the male, the question arises as to whether they may be due to the presence of the Y-chromosome. With relation to this the following is the situation:

1. In most cases, the presence of male secondary sex characters does not depend on the presence of the Y-chromosomes. This is shown by the following:
   a. There are animals which have no Y-chromosome, yet have male secondary sex characters: for example, the dog, the cat, the horse; also various insects.
   b. In Drosophila, males that contain no Y (as a result of non-disjunction) have the typical male secondary sex characters.
   c. In some breeds of sheep the male has horns, while the female has none. It could be suggested that this is due to the fact that

the male carries the Y-chromosome, while the female does not. But in other breeds, as Dorsets, both sexes have horns, although only the male has the T-chromosome. And in still other breeds, as Suffolks, neither sex has horns, although here as elsewhere the males have the Y-chromosome. If a female of the horned breed (Dorset) is crossed with a male of the hornless breed (Suffolk), in the Fi descendants the males have horns, although they receive their Y-chromosome from the hornless breed. They have inherited the horns through the mother. Thus it is clear that in these cases the presence of horns is not due to the presence of the Y-chromosome. Extensive work on the inheritance of horns and horn- lessness in sheep, goats and cattle consistently gives results which show that the presence of horns is not due to the presence of the Y-chromosome, or of any special type of Y- chromosome.

In all these cases (a), (b) and (c) the presence of distinctive male secondary sex characters depends in some way on the fact that the male carries but one X-chromosome, and consequently produces the male hormone. Their absence from the female results from the fact that she carries two X-chro- mosomes, and consequently produces the female hormone.

2. But in certain fish, some of the male secondary sex characters apparently do depend on the presence of a Y of a particular type, for they occur only where that type of Y is present, and they can therefore not be inherited through the mother. That is, the Y-chromosomes of certain individuals or breeds are so modified as to produce certain definite characteristics in the individuals that contain them, while individuals that contain another type of Y do not show these characteristics.

An example of this is found in the small fish Lebistes. In a certain race, the males have a black spot at the base of the dorsal fin, while the females are without the spot. In another race of the same species neither the males nor the females have the black spot. Thus the females of the two races are alike in lacking the black spot, while the males differ. Is this difference the result of differences in the Y-chromosomes of the two kinds of males?

This may be tested by crossing members of the two races. In the spotted race, represent the X and Y-chromosome by capital letters; while in the non-spotted race, represent them by the small letters x and y.

A female of the spotted race, XX, is mated with a male of the non-spotted race, xy, thus:

Females Males

Parents XX xy

Non-spotted
Fi Xx Xy
Non-spotted

THE Y-CHROMOSOMES

This produces in Fi females Xx, and males Xy. Neither contains the Y from the spotted race, and neither is spotted. The spot is not inherited through the females of the spotted race.

Next mate the female of the non-spotted race, xx, with a male from the spotted race, XY, as follows:

Females Males
Parents xx XY
Spotted
Fi xX xY
Spotted

This yields in Fi females Xx, and males xY. These males carry the Y from the spotted race, and they are spotted. Thus far the spot goes with the Y from the spotted race.

This Y may be traced further, for example, by breeding together the non-spotted females xx with the Fi males xY, from the preceding cross, thus:

Females Males
Parents xx xY
Spotted
F2 xx xY
Spotted

This gives in the Fs females xx, and males xY. These males are again spotted; they carry the Y from the spotted race. Wherever this Y from the spotted race is found, the black spot appears; wherever the Y is from the non-spotted race it does not appear. It appears clear therefore that whether the black spot appears in the male depends upon the type of Y- chromosome that it carries. There are two types of Y-chromo- some in the species, one producing the black spot, the other not producing it.

It has been shown by Winge that there are many other programmes markings in this fish that depend for their occurrence on the presence of a particular type of Y-chromosome. They occur only in the males, and only in males that have a certain type of Y. In certain other fish also the Y-chromosomes have been shown to be diverse in different individuals, giving rise to distinctive characteristics in the male.

In a certain plant, Melandrium, also, there are inherited characteristics that depend upon the presence of a particular type of Y-chromosome.

There are some indications also that in man the nature of the Y-chromosome affects certain characteristics. In a family described by Schofield, the male parent in the first observed generation had webbed toes. This was inherited in the male lines only by all the males for four successive generations. It thus followed exactly the distribution of the Y-chromosome from the original male parent. It seems probable therefore that it is due to a defect or modification in the Y-chromosome of that individual. Many other cases of webbed toes or fingers are known that occur in both sexes, so that they are clearly not due to the Y-chromosome.

On the whole, the situation as to the Y-chromosome and its relation to inheritance may be summarized as follows:

1. In most organisms Y is a degenerate chromosome, lacking most of the functions manifested by the other chromosomes.
2. In some organisms degeneration has gone so far that the Y-chromosome has completely disappeared.
3. In other cases it exists, but is very small.
4. But in some species the Y-chromosome has effects on hereditary characters that are similar to those produced by other chromosomes. Different types of inherited characters result in such cases from diversely modified Y-chromosomes.
5. Characteristics dependent on the presence of a particular type of Y-chromosome are inherited in the male line exclusively.

THREE MAIN METHODS OF DISTRIBUTION OF INHERITED CHARACTERISTICS

There exist therefore three diverse systems of distributing hereditary characteristics, depending respectively on the X- chromosomes, the autosomes, and the Y-chromosomes. In each of these the rules of inheritance are given by the special method of distribution of the three types of chromosomes, taken in connection with the facts of dominance and reces- siveness. Inheritance which follows the distribution of the X-chromosomes is commonly called sex-linked inheritance. That following the distribution of autosomes is known as autosomal or typical Mendelian inheritance. Less important than those just named is Y-chromosome inheritance. All these methods are included under the term Mendelian inheritance, employed in a broad sense.

In some organisms still other methods of inheritance exist. In higher animals and plants most inheritance occurs in one of the three methods named in the preceding paragraph.

4

The Genetic System as a Whole

Having examined the chief component parts of the genetic system, we proceed to an examination of the way it functions as a whole. This functioning and its consequences constitute what is usually called Heredity.

As we have seen, the genetic system consists mainly of the chromosomes with their genes: the X-chromosomes, the autosomes, and the Y-chromosome.

Why are the other parts of the germ cells not to be considered 'materials of heredity' or parts of the genetic system the cytoplasm as well as the chromosomes? It is known that the cytoplasm is of the very greatest importance in the development of the individual and in the production of its inherited characteristics. It is in the cytoplasm, through interaction with the genes, that the differentiations of the body arise.

The role of the cytoplasm, however, differs from that of the chromosomal materials in the following respects. Different germ cells differ effectively in their genes, and it is to these gene differences that the appearance of diverse characteristics in different individuals is due. But as a rule there is no evidence that the cytoplasm of different germ cells is effectively diverse in such a way as to produce different characteristics in different individuals. It is therefore the former the chromosomes with their genes that are commonly classified as the 'materials of heredity'more properly perhaps the 'materials of hereditary diversity'. The cytoplasm might correspondingly be called the 'materials of bodily differentiation.

Following the prevailing usage therefore, we may define the genetic system or material of heredity as follows:The genetic system consists of those parts of the cells that are effectively diverse in different germ cells, these diversities causing different characteristics to appear in the different individuals produced.

In rare cases, in certain plants, it is found that the cytoplasm contains programmes bodies which may differ in different germ cells and so cause differences in the characteristics of the individuals developed from them. In such cases the cytoplasm, or at least the programmes bodies it contains, must be included as part of the genetic system. But in most organisms the genetic system consists of the chromosomes with their genes.

CONSTITUTION OF THE CHROMOSOMES THE GROUPS OF CHARACTERS LINKAGE

Each of the chromosomes (save in some cases the Y-chromosome) is composed, as we have seen, of many diverse genes. In the wild, unmodified individuals, in the fruit-fly, the genes are such as produce the 'normal' or 'wild-type' characteristics of the individuals. The genes of any chromosome become modified in the course of time, in the process known as mutation, which will be dealt with in a later chapter. The modified genes produce altered characteristics, most of which are recessive, while the normal characteristics, when mated to the modified characteristics, are dominant. A few of the modified characteristics, however, are dominant when mated with the normal or wildtype characters. The modified characteristics resulting from mutation are in most, if not all, cases defects, as compared with the normal characters. Thus the normal, dominant characters form a standard condition, of which the defective and recessive mutated characters are modifications.

The diverse genes of any chromosome are designated, so far as they have received names, by the name of one of the modified characters which it helps to produce. Thus the gene yellow, near the upper or left end of the X-chromosome of Drosophila, produces in its mutated condition the modified body programmes yellow; while in its unmodified condition it produces the normal grey programmes of the wild fly. In the maps of all the names apply to the modified or mutated condition of the gene. For every such modified gene there is a corresponding unmodified normal or wild-type gene. It is the custom to designate the wild-type gene from which any modified gene is derived by writing with the name or abbreviation for the modified gene the sign +. Thus yellow + or y + or + y signifies the normal wild-type gene that produces grey body programmes; white + or w + signifies the normal gene located at I, i .5, which produces red eyes in place of white ones.

The genes of any one of the chromosomes, as X, or a given autosome, become modified in many different ways, so that different examples of the same chromosome have different sets of genes. The diversity of genes in the chromosomes of different germ cells causes diversities in development, with resulting different characteristics in the individuals produced. The characteristics produced by the action of the genes of any chromosome follow in the individuals of successive generations the distribution to the offspring of the genes that produce them. Thus many characteristics follow the course of the X-chromosome; many follow each autosome; a few may follow the Y-chromosome.

Since a large group of characteristics follows each chromosome, the result is that characteristics pass from parent to offspring in groups. As an X-chromosome passes from the father into his daughters, all the genes, and all the characteristics that they produce, pass together into the daughters. Here, of course, some of the recessive characteristics may remain hidden in the presence of dominant genes in the second X- chromosome carried by the daughters. But all may reappear anew when in a later generation this X-chromosome is again by itself in a male. Except in so far as there is exchange or crossing-over with the other chromosome of the pair, all the characteristics that depend on the genes of a particular chromosome pass together into the individual that receives that chromosome, and this continues for generation after generation.

Thus, in Drosophila, suppose that a male has white eyes and yellow body, two characteristics connected with genes in the single X-chromosome. He is mated with a female that has red eyes and grey body, again owing to genes in the X-chromosome. In the daughters of these two, the two types of X-chromosomes are present together. These daughters produce offspring in a third generation. Some of their sons receive the X-chromosome that came from the original male; these will have white eyes and yellow body. Other sons receive the X-chromosome that came from the original female; these have red eyes and grey body. Only a very minute proportion of the offspring have the reverse combinations of characteristics white eyes with grey body, and red eyes with yellow body. The two characters originally together white and yellow appear to be linked, and the same is true of the other combination, red and grey. The original combinations do not readily separate.This linkage of characters in heredity was observed before it was known that characteristics depend on chromosomes. Many theories were invented to account for it, until it was discovered that it is due to the fact that the linked characters depend on genes that are in the same chromosome. All characteristics that depend on the same chromosome form what is called a linkage group. Since there are different numbers of chromosomes in different organisms, there are correspondingly different numbers also of linkage groups.

These linkage groups are of great importance in heredity. They reveal to us many things about the nature and action of the genetic system. We shall therefore examine these groups in a typical organism, selecting for this purpose the animal in which they are best known, namely, Drosophila melanogaster.

THE GROUPS OF LINKED CHARACTERS IN DROSOPHILA

The fruit-fly has, as we have seen, four pairs of chromosomes. On the genes in these four pairs of chromossomes depend several hundred known inherited characteristics. These many characteristics form four linked groups, according to their dependence on genes in the X-chromo- some, or in the chromosomes II, III or IV. It is to be noted that all of the characteristics that depend on genes in either one of a given pair of chromosomes constitute a single linkage group, since all those characteristics may become connected with the genes of a single one of the chromosomes. Thus the number of linkage groups is the same as the number of pairs of chromosomes. In the course of breeding successive generations, linkage of charact-eristics shows itself in the fact that certain characteristics (dependent on genes in the same chromosome), which are present together in one parent, are present together also in the individual descendants. This is best seen on comparing grandchildren with grandparents. In this case one grandparent had the combination white-yellow, the other the combination red-grey. The same two combinations reappear in the grandchildren (seen particularly in the grandsons). In addition there are a very few 'cross-overs', having the new combinations white-grey in some individuals, red-yellow in others. But almost all the grandchildren get both the characteristics from the same gandparent. If the original grandparents have the other combinations, grey-white and yellow-red, then the majority of the grandchildren show these combinations.

Thus just what combination of characteristics the individual grandchildren have depends on what characteristics were united in the grandparents. But this is the case of course only for characteristics whose genes lie in the same chromosome. If two characteristics depend on genes that are in different pairs of chromosomes, then the combinations that were present in the grandparents are no more frequent in the descendants than are the reverse combinations.

It is in many respects most convenient to think of linkage in terms of the gametes. The combination of genes that are present in the gametes can often be determined by examination of the offspring produced. With relation to the gametes, linkage can be characterized as follows. When an individual (zygote) is formed by the union of two gametes, each bearing in one of its chromosomes a certain combination of linked genes, this same individual later produces gametes of which

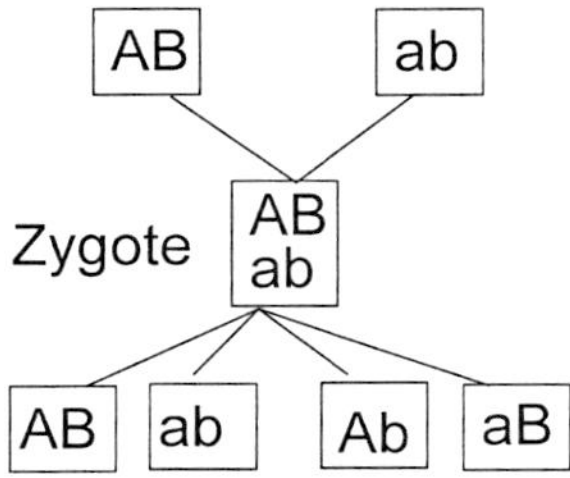

• Gametes • Gametes • Mary Few

Fig. Diagram of the results of linkage, with relation to the gametes. The two gametes above carry respectively the linked genes AB and ab. The zygote produced by their union carries both AB (in one chromosome) and ab (in the other). Then the majority of the gametes formed by this zygote carry again the linked genes AB or ab; only a minority cany Ab or aB.

the majority have these same combinations of linked genes. Often also there is a minority of gametes that have the genes combined in a new way, as a result of crossing-over. In the fruit-fly, crossing-over occurs only in the females, not in the males. Thus if a male individual is formed by the union of two gametes carrying certain combinations of linked genes, then later such a male produces only gametes having the same combinations of linked genes. This fact is extremely convenient for determining what characteristics are linked in Drosophila.There are many other organisms in which crossingover is not limited to one sex. By breeding for successive generations and by the study of exchanges between the chromosomes, the following groups of linked characters have been found in the fruit-fly.

Group IA large group of sex-linked characters, more than 100 in number.These all follow in heredity the distribution of X-chromosomes, as described in earlier pages. Any two or more of these characters may be linked together and connected with the same X-chromosome.

Group IIA large group of typical Mendelian characters, 50 to 100 or more in number. These follow the method of distribution of autosomes, and any two or more of the group may be connected with the same autosome.

Group III.Another large group of typical Mendelian characters, 50 to 100 or more in number, following the method of distribution of autosomes. Any two or more of this group may be connected with the same autosome, but a character of Group III is not linked with any character of Group II.

Group IV.A small group of typical Mendelian characters, only 5 or 6 known.These are not linked with any of the characters of Groups I, II or III.All of the 300 or more characters known in this species fall into one or the other of these four groups. The nature of the characters found in the different groups is considered on later pages.

Thus here the number of linked groups is the same as the number of pairs of chromosomes. There are three large groups and one very small one; likewise there are three large pairs of chromosomes and one very small pair. It is natural to conclude, therefore, that the groups of linked characters correspond to the pairs of chromosomes, and that the characteristics in each group are connected with some particular pair. As it turns out, there is conclusive evidence that this is true. The evidence for each of the groups may be summarized as follows:

All the characters in Group I are sex-linked; that is, they follow exactly the distribution of the X-chromosomes. Any character in this group follows a particular X-chromosome and its descendants, wherever that chromosome goes. The characteristics in Group I are thus certainly dependent on genes in the X-chromosomes.

The small group of characters known as Group IV has been proved to be connected with the small chromosomes of pair IV. This has been demonstrated in the following manner. By non-disjunction of this fourth pair of chromosomes, some individuals are produced having but one small chromosome (IV). That this is the situation can be seen under the microscope, in properly prepared material. These individuals with but one of the small chromosomes have the dominant normal or 'wild' characteristics. Such individuals are mated with individuals containing one of the recessive characters of Group IV. One of these recessive characters is 'bent wings'; another is 'eyeless' the individual having no eyes.Half of the germ cells from the individuals that have but one of the fourth chromosomes are without this chromosome, while half of them have it. These two kinds each unite with germ cells carrying the recessive character 'bent', or 'eyeless'. There are thus produced individuals, half of which have but one of the fourth chromosomes, while the other half have two. In those which contain two, all the characteristics are of the dominant normal type; the recessive character bent or eyeless is not manifested. But in the individuals having but one fourth chromosome, the recessive character is manifested: the individual shows bent wings, or is eyeless. This demonstrates that the characteristics of Group IV are connected with the small chromosomes.

Thus the large linkage Groups II and III are left for the two large V-shaped chromosome pairs, II and III. These chromosomes can be broken into two or more pieces by subjecting them to radiations. When this is done, the characters of the corresponding linkage groups are no longer linked together.Thus in Group III the two genes scarlet and sooty are linked (completely in the male). But when the corresponding large chromosome (III) is broken into two pieces, scarlet and sooty are no longer linked; they may pass to different offspring.

They are evidently connected with the two separated pieces of chromosome III. In the same way the other large chromosome (II) may be broken, and this causes the characteristics of the linkage Group II to separate into two groups. Minute study shows that the chromosomes II and III differ a little, in size and form, so that with practice it becomes possible to distinguish one from the other under the microscope.Thus it is proved that each of the four linkage groups is connected with one of the four chromosome pairs; the four groups correspond to the four chromosomes.In a number of other organisms the number of linked groups of characters has been determined, and also the number of pairs of chromosomes. In every case in which this has been fully worked out, the number of linked groups is the same as the number of pairs of chromosomes. The numbers of groups and of pairs in certain organisms are as follows:

Drosophila melanogaster 4, Drosophila virilis 6, garden pea 8, sweet pea 8, maize10.In man there are twenty-four pairs of chromosomes, including the X's. It is to be expected therefore that there will be found 24 groups of linked characters, one group sex-linked, the other 23 groups typical Mendelian, in their inheritance. But it will be a very long time before these 24 groups are identified.

FUNCTIONS OF DIFFERENT CHROMOSOMES IN RELATIONS TO CHARACTERISTICS

Have the different chromosomes different functions? Can we classify their functions in any simple or systematic way? Do certain chromosomes influence certain particular parts of the body, or certain particular functions, while others influence other parts or functions?

Such questions can be answered by an examination of the kinds of characteristics found in each of the linked groups; this will show what kinds of characteristics are influenced by each of the chromosomes. We give, therefore, the following lists of a number of typical characteristics connected with each chromosome in the fruit-fly. The characteristics listed are all modifications of the normal characteristics found in the typical wild fly. Hence for every characteristic here listed there exists also a corresponding unmodified normal characteristic. Each of the characters listed is the result of modification of a single gene.

List of some Typical Characters connected with the Genes of each of the Four Chromosomes of the Fruit-fly

In the lists, the part of the body, or the function, affected by the character is first indicated.This is followed by the names that have been applied to the different characteristics. Many of these are self-explanatory; of some others a brief characterization is given.

Group X-chromosome.Sex-linked inheritance.Eyecolour: coral, blood, eosin, cherry, apricot, buff, tinged, ivory, white (the foregoing are multiple alleles; that is, they are all diverse modifications of the same gene). Modifications of other genes: vermilion, ruby, carmine, garnet.

Eye form or structure: bar, facet, furrowed.

Body colours or markings: yellow, sable, lemon, green, chrome, tan, spot, dot, etc. Wings: rudimentary, miniature, bifid, bow, notch, depressed, club, fused, cut, etc.

Legs: reduplicated.

Body structure: abnormal abdomen.

Bristles: forked, singed, tiny, bobbed, scute.

Reactions to light: tan (individuals showing this programmes characteristic do not fly Towards the source of light, as do the normal individuals).

Life and health: at least 10 different lethal genes are known to be connected with the X-chromosome. Any one of these prevents the animal from developing, if no corresponding normal gene of the same pair is present. All, or practically all, the characteristics in the above list cause the organism showing them to be weaker and less resistant than the normal 'wild-type' flies.

As will be seen from the above list, the genes of the X-chromosome affect practically all parts of the body, and many of its functions.

Group II. Second chromosome (autosome). Mendelian inheritance.

Eye programmes or structure: purple, pinkish, cream, star, morula, etc.

Body programmes and pattern: olive, black, streak (thorax), comma, trefoil, speck, patched abdomen, etc.

Wings: truncate, balloon, vestigial, blistered, jaunty, curved, strap, apterous (and many others).

Bristles: fringed, minute, vortex.

Life and health: lethals (several).

Thus chromosome II affects many parts of the body.

Group III. Third chromosome (autosome).

Mendelian inheritance.

Eye programmes or structure: claret, cream III, deformed, kidney, maroon, pink, peach, rough, scarlet, sepia, etc.

Body programmes or pattern: band (thorax), ebony, sooty, trident, etc.

Body structure: bithorax, dwarf, giant.

Wings: beaded, spread, tilt, truncated, etc.

Bristles: doubled, hairless, hairy, minute, spineless, two- bristle, vortex III, etc.

Life and health: several lethals.

Thus chromosome III also affects many parts and functions.

Group IV. Fourth chromosome (autosome).

Mendelian inheritance.

Eyes: eyeless.

Wings: bent; cubitus interruptus.

Bristles:shaven.

Few genes known, but affecting diverse parts of the body.

What does the foregoing list of characters affected by the different chromosomes show as to the nature and functions of the chromosomes?

1. All the large chromosomes (Nos. I, II, and III) affect all parts of the body: many different functions, health, length of life, and the like. The small pair (IV), though affecting few characters, still affects diverse parts (eyes, wings, bristles).
2. Thus particular chromosomes are not limited in their action to any particular part of the body or to any particular function. Each chromosome affects many parts and functions.
3. It appears clear that each chromosome enters into the process of development, affecting many processes, and doubtless influencing the entire organism. The method of action of the X-chromosome has been treated in a previous chapter; as there seen, it alters the chemical processes, thus affecting the development of all parts of the body.The other chromosomes doubtless act in a somewhat similar manner.
4. It is not clear from our present knowledge that the genes are arranged in the chromosomes in any systematic way, with relation to their functions, or to the parts of the body that they chiefly influence. A gene principally affecting eye programmes is close to one principally affecting body structure or body programmes or structure of the

wings, and so on. Genes affecting eye programmes occur in many different parts of the genetic system. In a similar way, genes that affect the structure of the wing or of the eye, or that affect the programmes of the body, are scattered throughout the four chromosomes.

Thus if there is any functional system in the arrangement of the genes in the chromosomes, it has not yet been discovered; in the present state of our knowledge it appears as if there were none.

By 'characteristics' or 'characters' are meant any features, physical, chemical, physiological, or structural, in which organisms may differ. The materials presented in the chapters preceding this one have brought out, explicitly or implicitly, many of the relations between genes and characteristics. Here these relations, and others not yet set forth, are brought together in systematic order.

1. SPECIAL ACTION OF A GENE

Particular genes affect mainly, though not exclusively, some particular part of the body, or some particular function.

Thus, in Drosophila there is in the X-chromosome, at the locus I, 1.5, a certain gene which in its normal condition takes part in producing the red eye programmes. This gene has become modified in different cases in a considerable number of different ways, and any modification alters the programmes of the eye. The different modifications give the many different alleles listed earlier among the eye-programmes characters of linkage Group I. Other genes have their chief effect on the body programmes or on the wings or on some other part. Particular genes in man affect the blood type of the individual; others affect the eye, the body form, and so on. Each gene as a rule has its main function in connection with the development of some particular part.

2. Yet a given gene often produces effects on several parts or functions. For modification of a single gene may cause a change in the characteristics of a number of different parts. Some examples of this may be given.

The character bent wings, belonging to linkage Group IV of Drosophila, is due to the modification of a single gene in chromosome IV. It causes the wings to be bent dorsally at their base, and also it causes the legs to be shorter than normal, owing to the abnormal shortness and thickness of the meta- tarsal joint.

In animals showing the character vestigial wings (due to modification of a gene in chromosome II), the following peculiarities appear: the wings are degenerate, only their bases being present, and these are held at right angles to the body. The balancers are likewise degenerate. Two rear bristles on the scutellum are farther apart than usual, and they stand up straight instead of inclining backward. The flies hatch about two days later than in the normal wild flies. They are not quite so vigorous as the normals.

As noted in the preceding chapter in connection with linkage Group I, modification of a certain gene in the X-chromo- some causes;

a. The body programmes to be tan instead of grey,
b. Causes the individuals to lose the usual positive reaction to light.

In any detailed account of the effects of particular genes, it will usually be found that each gene has thus several different effects, often on diverse parts of the body. The effect on which the name given to the gene is based is commonly merely that effect that is most conspicuous to the eye.

3. General or Constitutional Effect of a Gene

In addition to the special effects on certain structures or functions, modification of any gene has also a general effect on the bodily constitution. In Drosophila the 'wild-type' flies, with unmodified genes, are vigorous and long-lived. Practically all modifications of the normal genes make the animals less vigorous, so that individuals showing any of the characteristics listed in the preceding chapter are weaker than the normal or 'wild- type* individuals, and have a higher rate of mortality. Thus practically all the modifications of normal genes are defects, a fact to which we return in a later chapter.

In view of the facts set forth in preceding paragraphs it has been said with much probability that 'every gene affects the development of the entire body.

4. Thus apparently each gene supplies material which in the process of development enters into reaction with materials from all the other genes; it thus modifies many or all parts of the body, but affects certain parts more conspicuously than others.
5. As before set forth, a particular single gene may become modified in many different ways, in different individuals or different chromosomes, each modification causing a changed characteristic. The classical example of this is the series of different eye colours (multiple alleles) found in different individuals of Drosophila, each programmes being due to a different modification of the gene located at the point I, 1.5 in the X-chromosome. This gene in its normal condition produces (by interaction with other eye-programmes genes) the normal red programmes. The various modifications of the gene give rise to the colours before mentioned: coral, blood, eosin, cherry, apricot, buff, tinged, ivory, white.

Many such series of multiple alleles are known in the fruit- fly and in other organisms. In the fruit-fly the following illustrate their nature:

Scute: The gene scute is close to the upper (or left) end of the X-chromosome. It exists in 20 to 30 different modifications, each of which causes a slight change in the distribution of the bristles on the body.

Cut: A considerable number of different alleles are known of the cut gene, located at I, 20.0. All of them cause changes in the outline of the wings, as if bits of the wing were cut off.

Vestigial: The gene vestigial at II, 67, which reduces the size and changes the form of the wing, has two other alleles, antlered and strap> giving wings of other forms and sizes.

Garnet: At the locus I, 44.4 are several alleles which change the programmes of the eye in diverse ways. Some of these are known as garnet, salmon, garnet 2, garnet 3.

Pink: The eye colours, pink, peach, pink 3, pink 4, pink 5, are a series of alleles located at III, 48.0.

Series of alleles are known also in other organisms. Thus in the rodents (rats, guineapigs, rabbits), series of different alleles are known, yielding different colours of the hairy coat. In maize there are several allelic series modifying the programmes of parts of the plant. Presumably any gene may, by successive modifications, produce a series of alleles.

As mentioned in an earlier chapter, all the members of an allelic series affect the same character or part of the body, yielding different modifications of that character. Further, they show certain peculiarities of inheritance as compared with different genes that are not alleles. Since they are all modifications of the same gene, only one of the alleles can be present in a single chromosome, and for this reason only one of them can be carried by a single gamete (except in cases of non-disjunction). Further, only two members of a given allelic series can be present in any diploid individual or zygote.

There are thus two types of diversities among genes,

a. On the one hand, there are the different genes occurring at different loci in the chromosomes, the diverse genes, shown on the maps affecting many diverse characteristics.
b. On the other hand, there are these different modifications of a single one of these genes the multiple alleles.

6. Any given single character, inherited in the sex-linked or the autosomal manner, is a product of the interaction of many genes, located in different parts of the genetic system.

As an example of this, the programmes of the eye in Drosophila may be taken. As shown in the preceding chapter, the programmes of the eye depends on many genes scattered through the different chromosomes. The number of different genes directly affecting the programmes of the eye is indeed still greater than this, and in addition there are many other genes that affect eye programmes indirectly, through the fact that they affect the structure of the eye: for example, the gene 'eyeless* in the fourth chromosome pair. If this gene is modified in a certain way, no eyes are developed, consequently no programmes.

Similarly, many different genes affect the form and structure of the wings, the venation of the wings, the structure of the eye, the programmes of the body, the form of the body, the structure and distribution of the bristles. Indeed, this is presumably the situation with relation to any characteristic whatever; every characteristic is influenced by many genes.

7. Altering any one of the numerous genes that cooperate to produce a given characteristic alters that characteristic. Any characteristic may, therefore, be changed by altering any one or more of many different genes.

Thus, consider what must be the situation among the genes to yield the normal red eye programmes in Drosophila. To produce this result all of the 19 (and more) genes that affect eye programmes must be present in the usual or normal condition. If any one of them is changed (the rest remaining unchanged), the programmes of the eye is altered. Changing in a certain way a gene at I, 1.5 all other genes remaining normal, the eye programmes is changed to white. If we change the gene at I, 33 (the rest remaining normal), the programmes becomes vermilion. If we change the gene at II, 54.5 (the rest remaining normal), the programmes of the eye becomes purple. Thus one could go over the entire 19 genes; changing one after the other; each one changed alters the eye programmes.

Again, consider what must be the situation in the genetic system to yield one of the recessive colours: for example, vermilion eye programmes. To produce this programmes, 18 of the genes must be in their usual normal condition, while a single one, at the point I, 33, must be altered. If, in addition to this one, another is altered (for example that which by itself yields purple, at II, 54.5), a different programmes is produced. Altering one gene changes the programmes in a certain way; altering two or more changes the programmes in a still different way. Similar relations would be found with respect to most or all characteristics.

8. Multiple Factors

The different genes on which a given character depends may have qualitatively different effects; such is the case with the different genes that affect the eye programmes in the fruit-fly. In other cases the different genes affect the characteristic only quantitatively. Thus dimensions often depend on many different genes. Modifying one of these or substituting another for it changes the dimension; modifying an additional one changes it still more, and so on. In such cases the character is commonly said to depend on 'multiple factors'.

9. 'Duplicate Factors

   In some cases a character may depend on the presence of two (or more) genes of a particular type in such a way that both of these genes must be present in order that the character shall be produced. The character is then said to depend on duplicate genes or factors. In the plant known as the shepherd's purse, the common type of pod is triangular, but some plants have an elongated oval pod. This oval pod occurs only when two particular recessive genes are present in the homozygous condition. Thus if the two dominant genes which produce the triangular pod are A and B, while the two recessives that give the oval pod are a and b, the oval pod occurs only in the plants having the genes aabb, while all the other combinations give the triangular pod. Similar cases of duplicate genes are known in a number of organisms.

10. 'Modifying Factors'

    a. In some cases the effect of some of the genes is merely to modify in detail a characteristic that is due primarily to another gene. Such are known as modifying genes or factors. Thus, in Drosophila, a certain modified gene at I, 1.5 causes the eye to assume the pinkish programmes known as eosin, but the depth of this eosin programmes depends upon certain other genes present in other parts of the genetic system. One of these, known as cream b, is located at II, 22.5; another, cream III, at III, 36.5; and several other such genes modifying the eosin programmes are known. If these genes are present in the normal or unmodified condition they do not affect the eosin programmes; when modified in certain ways they do. Again, in the rat, the presence of a certain gene causes the fore part of the body to be coloured brown, forming what is called a hood) the rest of the body being white. The extent of this hood depends on the presence in the individual of a number of other genes; some of them increase its extent, others decrease it. Such modifying genes or 'modifying factors' play a role in many characteristics. They may produce no visible effect except in the presence of a particular character that they modify. Thus the modifiers of eosin eye programmes produce no visible effect in an individual with red eyes.

b. The same phaenotypic character that is, the same characteristic in outward appearance may be due in different cases to the alteration of diverse genes, in different parts of the genetic system. This is illustrated in numerous cases in Drosophila.

Thus, modification of a certain gene at II, 54.5 causes the programmes of the eye to change from the usual red to a darkish purple; the modified gene giving this programmes is known as purple. But in other cases what is outwardly the same programmes is found to be produced by change in other genes. One of these is at I, 44.4; to distinguish it from the one just mentioned it is called garnet. A third gene that produces practically the same eye programmes is found at III, 49.7; it is called maroon.Similarly the eye programmes 'salmon', resulting from a modified gene at I (again at 44.4), is indistinguishable from the eye programmes 'rose', resulting from a gene modification at III, 48.0. Other cases of this sort could be given. In such cases the same phaenotypic character will be found to be inherited in different ways in different cases, depending on the location of the modified gene to which the character is due. If the gene is in X, the inheritance will show itself to be of the sex-linked type; if in II, it will be autosomal and of the linkage Group II; if in III, it will be autosomal and of the linkage Group III.

THE ACTION OF GENES IN DEVELOPMENT

In view of what has just been set forth, how are we to conceive the operation of the genes in producing the characteristics? How do they operate in producing an adult from the fertilized egg? And how do different sets of genes produce individuals having different characteristics?

On the intimate details of the operation of the genes little is known. Some of the main features in their method of action were taken up earlier in dealing with the relation of chromosomes to sex, though only the action of large groups, forming chromosomes, was there dealt with. From the relations brought out above, and in preceding chapters, taken in connection with the facts of embryological development, it is possible to form a conception of some of the main features in the action of genes in building up the body. These are summarized in the following.The genes form a set of many diverse materials, differing from each other chemically and physically and having different chemical and physiological activities. They exist at the beginning of the individual's development in the nucleus of the fertilized egg, while surrounding them is a mass of cytoplasm.The genes interact with the cytoplasm, changing it, producing new materials that are incorporated into the cytoplasm. These varied cytoplasmic products of the different genes interact with each other, producing again new types of material.

Some of the gross features of the interaction of genes and cytoplasm were described earlier. As there shown, the chromosomes (groups of genes) visibly take up material from the cytoplasm, alter it, and give it off again into the cytoplasm. Though the genes thus interact with the cytoplasm, modifying it, they are not themselves used up in this process so as to lose their essential character. They supply material which combines with cytoplasmic material. But there is left an unchanged part of each gene, and this by division produces additional genes of the same type and action.

As the cells divide, each chromosome divides, and each of its contained genes divides. One of the two halves of each gene goes after division to each of two cells produced. Thus each new cell produced contains the full set of genes, all those that were present in the fertilized egg.

In most organisms it appears probable that this continues throughout at least all the early stages of development, if not to the end of development, so that all the cells contain all of the genes. For in the young embryo, after many hundred cells have been produced, it can be shown that any cell can produce many different parts of the organism. And in many organisms even in late stages there are cells that can produce any part, or the whole organism, as is shown by the facts of regeneration. Such cells must contain the entire set of genes.

In very late stages of development it is possible that in highly differentiated tissue cells some or all of the genes have become lost or modified, though there is no clear proof of this. Further, there are a few organisms of which it is to be observed that the cells that are to produce the differentiated body of the animal lose parts of their chromosomes, and so presumably some of the genes. The chromosomes in these cells are seen to break up, and portions of them are dissolved and absorbed by the cytoplasm while in the germ cells the chromosomes remain entire. This is known to occur in threadworms (Nematodes) and in certain insects. The relation of these processes to the way bodily development occurs is quite unknown.

But on the whole the evidence is strong that in most organisms the production of different tissues and organs by different cells is not due to their containing different genes, for cells that certainly contain all the genes produce many different parts in many different cases. The genes act differently in the different cells, depending on the relation of the different cells to each other, so that the different cells, though containing the same genes, produce different tissues and organs.

The diversity of the different parts of the body seemingly arises mainly, if not entirely, through the fact that different cytoplasmic products of the genes are separated into the different cells. Thus the cells become different in their cytoplasm. And this different cytoplasmic constitution of the different cells interacts differently in each case with the full set of genes that the nucleus contains, so giving anew different products. By a continuation of such action the different tissues and organs are produced. In each, the cells contain the same sets of genes, but differ in their cytoplasm. The differentiations of the body therefore lie mainly in the cytoplasm, though the diversities in the cytoplasm are brought into existence through the action of the genes.

Yet it is true that differences may arise within the body of a single individual through a change in the genes of some of the cells. As before seen if in a female Drosophila one of the cells loses one of the X-chromosomes, that part of the body derived from this cell shows the characteristics of the male, while the rest is female. If the X that is lost carries dominant genes, while the remaining X has the corresponding recessive genes, then, in that part of the body whose cells have lost one X-chromosome, the recessive characters are manifested, while the remainder of the body is dominant. In this way are produced mosaic organisms, showing diverse hereditary characteristics in different parts of the body. Such mosaic diversities, however, appear to be of a very different type from the differentiation of the body into tissues and organs. ^ Whether any differentiations in ordinary development are thus produced through loss or modifications of certain genes must be considered as yet uncertain, though it is entirely clear that most of them are not so produced.

But diversities between individuals, as we have seen, commonly result from differences in the genes that they contain. If two fertilized eggs begin with different sets of genes, the processes of changing the cytoplasm, of differentiating it into tissues and organs, differ in the two cases, so that individuals with diverse characteristics are produced. Very marked differences in characteristics may be produced even when the two sets of genes differ in but a single gene or pair of genes. In Drosophila, as before seen, if two eggs differ in a single pair of genes in chromosome IV, one of them produces individuals without eyes, while the other has normal eyes. If two individuals differ in an entire chromosome (X), one becomes a male, the other a female.

Such differences are due to the diverse chemical and physical action of the different genes. They produce differences among different cells, in the way set forth in earlier paragraphs. Also, they may produce diverse chemicals (hormones) which circulate through the body, thus causing changes in the development of parts with which they come in contact.

Different chemical and physiological effects, different development, may be produced by difference in balance among the genes. Two cells may contain the same kinds of materials (genes) throughout, but one of these contains certain materials in two doses (a pair of genes), while the other contains but one dose (an unpaired gene). In consequence the two cells differ greatly in development. In this way difference in sex is produced, with its manifold attendant differences in structures and functions. In many cases two genes of a certain kind produce a different effect from a single one, though in other cases they do not. That is, two centres of a certain kind of action may give different results from a single centre. This principle of the effects of difference in balance is presumably of great importance, though as yet not a very great deal is known as to its operation.

It appears possible that the operation of some or all of the genes begins in early stages of development and continues throughout. It is known that some genes produce their distinctive effects at the beginning of development, in the egg itself. Thus in the silkworm a difference of one gene is known to make a difference in the programmes of the egg. Two races differ in egg programmes, one having slate-coloured eggs, the other brown eggs. Crossing the two races shows that the difference in programmes is due to a difference in a single gene, for the descendants show 'unit difference' inheritance. If, by crossing, the special gene from the slate-coloured race is brought into the eggs of the brown race, in the next generation the eggs are slate-coloured instead of brown, so that the 'slate' gene is active in the formation of the egg.

Other genes are known to produce their distinctive effects somewhat later, but still rather early in development. Thus in mice, if one of the genes of a certain pair is modified in a certain way, this causes the hair of the individual to be yellow. If both the genes of this pair are so modified as to produce the yellow programmes, the individual develops for but a short time, then dies while still an embryo. It is clear that these genes are active in the young embryo. Again, in plants, certain genes are required for the production of chlorophyll. If one of these becomes defective, chlorophyll is not produced. The young plant grows so long as the stored food in the seed is available, but, having no chlorophyll, it cannot make use of the sun's rays for the elaboration of nutrition; it therefore dies as a young seedling. Examples of the operation of genes during early development are abundant.

Some genes are required for laying the foundation of certain parts or organs; if these genes are altered, no such organ is produced. Thus in the fruit-fly a certain gene in chromosome IV is necessary, in its normal condition, for producing eyes. If that gene becomes defective, the eyes are completely lacking, although all the other genes that cooperate to produce the eye may be present in normal condition. This particular gene is required for the very foundations of the eye. Similarly, in vertebrates a certain gene is required for the production of the pigment which produces the colours of hair and skin. Many different genes cooperate to produce this pigment; most of them modify the type of pigment produced. But if this particular one becomes defective in a certain way, no pigment is produced; the animal is an albino.

A somewhat similar situation appears to be at the basis of some cases of feeblemindedness in man. In such cases the brain is not properly constituted; it does not function well. Study of inheritance shows that in some cases, at least, feeblemindedness is due to a defect in a single gene; there is 'unit difference inheritance. This single gene plays a necessary part in producing an efficient brain; when the gene is defective the brain is fundamentally imperfect.The special and distinctive functions of some genes are not known to come into operation until late stages of development. Such, for example, are those that produce in Drosophila the distinctive pigments of the eye. Their visible effect does not appear until the eye is formed; although of course this action may have begun early in development.

In many such cases in which the conspicuous effect of a gene appears at a relatively late stage in development there are less conspicuous effects that indicate that the gene may be in operation throughout development. Thus individuals with white eyes, or with rudimentary or vestigial wings, are less vigorous than those with normal eyes or wings; they are less resistant and have shorter lives, and this is true for most of the types having recessive characters. Though the conspicuous special effect of the changed gene is late in appearing, the effect on the general constitution indicates that there may be an effect throughout development.

That this is indeed the case is indicated strongly by certain other facts. Sometimes a gene at a certain locus in the chromosome becomes entirely inoperative; it has either dropped out of the chromosome, or it has totally lost its action. Such cases are commonly known as deficiencies. Deficiencies are producible by the action of radiation on the chromosomes, so that large numbers of them have been studied. A deficiency is detected in the following manner. A chromosome suspected of being deficient for a certain gene at a certain locus is brought into an individual in which the other member of the pair to which it belongs carries a recessive gene at that locus. If the chromosome is deficient in the suspected gene, the recessive gene is manifested, since there is present no dominant gene of that pair. Thus if an X-chro- mosome carrying white eye is brought into a female whose other X is deficient for the gene at the white eye locus (the locus 1.5), the female has white eyes.

In the case of almost any gene, if two chromosomes that are deficient for the same single gene are brought into the same fertilized egg, so that the egg lacks that gene entirely, the egg does not fully develop. In most cases it is not known at just what stage development ceases. But the fact shows that, in addition to its special effect on a particular characteristic, the gene is necessary if development is to continue at all beyond a certain stage. By ingenious methods Demerec has recently proved that if only a few of the cells of a developing embryo are deficient for a certain gene, those few cells fail to develop, though the rest of the body 'does develop.

This was demonstrated for deficiencies of many different genes in the X-chro- mosome of Drosophila. In only one of the many cases tested was deficiency for a certain part found not to be lethal, though even in this case it caused weakness. In the case of the white eye gene at I, 1.5, and in many other loci, deficiency prevented development. Altogether, it is clear that genes are active throughout development, some beginning early, others seemingly later, and that almost all genes are necessary in order that full development shall occur. Changing one or more genes may cause from the beginning a change in the chemistry and physiology of the organism, altering its course of development and so giving rise to changed characteristics.

WHAT KINDS OF CHARACTERISTICS ARE AFFECTED BY GENES?

The question as to what kinds of characteristics are affected by genes has been much discussed. Some have supposed that only a few characteristics depend on genes: only certain superficial matters, such as colours, and the forms of external features. Particularly with relation to man has there been dispute on this matter.

But on this question, positive knowledge is available, both for man and for other organisms. It is possible to answer fully the question: What kinds of characteristics are affected by genes?

The question can mean only: What kinds of characteristics can be altered by changing the genes? What kinds of differences in characteristics result from differences in genes?

There are several ways of answering these questions. The most direct is by the crossing of individuals having different genes. This results in substituting one type of gene for another, with consequent differences in characteristics. If the diversities in characteristics show Mendelian or sex-linked inheritance, this demonstrates, as we have seen, that they are due to differences in genes.

By this test, in many organisms (for example Drosophila), the following kinds of characters have been found to be affected by genes:

1. Structures and forms of all parts (head, eyes, thorax, abdomen, wings, legs, internal organs).
2. Colours and patterns of all parts. These depend on chemical processes, hence:
3. Chemical processes of many kinds are affected by genes.
4. Physiological conditions: health, vigour, life and death.
5. Power of development: certain lethal genes stop development at a certain stage. A defective gene in the fourth chromosome of Drosophila prevents the development of the eye. Certain genes cause the body to grow to a larger size than usual. Many examples of the effects of particular genes on development have been given in earlier pages.
6. Behaviour. Reactions to light in Drosophila are changed by alteration of a certain gene ('tan').
7. Sex. Changing the genes changes the sex; this involves changes in all types of characteristics: in structure, pattern, programmes, physiology, behaviour.

Thus, in sum, it is clear that in animals, characteristics of all kinds, without exception, are affected by genes. There are no characteristics that are exempt from influence by the genes. Any characteristic can be altered by suitable alteration of genes. This is not surprising when it is considered that the genes are among the materials from which organisms are made, and that altering them alters the chemistry of the developmental processes.

What is the situation in Man? What characteristics are affected by genes in Man?

In man we have to deal with certain important types of characteristics that are hardly to be studied in other organisms. These are what we call mental characteristics: emotions, temperament, skill, character, intelligence, reasoning power, and the like. There are great differences among human beings in their mental characteristics, and in their capabilities. Are these things affected by genes? Often it is argued that they are not. Some have held that practically all differences in mind and character are due to environment, to education, influence of associates, tradition, state of civilization, and the like.

On this matter there is positive knowledge which completely removes the matter from doubt. There are two methods by which it can be determined in man what kinds of characteristics depend on genes.

1. The first method is by observing what characteristics show Mendelian or sex-linked inheritance. This method is not entirely satisfactory in man, since;
   a. It is not possible to carry on experimental breeding in man,
   b. This method can hardly be applied to the important mental characteristics. Yet by this method many characteristics of man have been shown to depend on genes. Among the characteristics showing sex-linked or Mendelian inheritance, and so depending on genes, are the following in man:
   i. Programmes of hair and eyes; form of many facial features; structure of the hands and feet; form of the body, whether stout or slender; chemical constitution of the blood; efficiency of the brain (whether the individual is normal or feebleminded); condition of the eyes and sense of sight (colourblindness, short-sight, night-blindness and other conditions); sense of hearing (deafness, etc.); sex, and many other characteristics.
2. A second method gives much fuller and more detailed knowledge of what characters are affected by genes in man. This is by the study of identical twins, as compared with other individuals. This is a most important method, and it will therefore be discussed in some detail. It depends on the folio wing facts:
   a. In man, most individuals are derived from separate germ cells, and therefore have diverse sets of genes. For in forming germ

cells, as before described, reduction of chromosomes and genes occurs, so that practically every germ cell differs in its gene combination from every other. When by fertilization two different germ cells unite, the diversity of gene combination is increased still more. The result is that no two individuals formed in this way have the same combination of genes.

b. But rarely there occur cases in which a single fertilized egg produces two individuals (twins). Such individuals are formed by division of all the genes, so that the two twins contain the same gene combination. These are the twins known as monozygotic or identical twins. In another type of twins, the two are derived from separate eggs; such dizygotic twins have diverse gene combinations.

By comparing the similarities and differences of identical twins with those in other individuals, and particularly with those of twins of the other type, it is possible to find out what characteristics are affected;

1. By having all the genes alike,
2. By having different combinations of genes.

'Identical twins', or offspring all having the same genes, occur in many organisms. In lower animals (Protozoa, plants, Hydra, worms, etc.), a single individual often divides into two or more, by Vegetative* reproduction. All these individuals have the same sets of genes, since in vegetative reproduction every chromosome and every gene divides. Such individuals are very exactly alike. They have been studied a great deal. In such a Protozoan as Paramecium, a great number of individuals are produced in this way, all closely alike. But when conjugation takes place, this gives new gene combinations, and the individuals after they have conjugated are very different.

In one of the vertebrates, the armadillo, identical twins are formed in essentially the same way that they are in man, but are formed regularly. A single fertilized egg, after developing some distance, divides into four, so that four 'identical twins', or quadruplets, are produced at each birth. The four are extraordinarily alike even in fine details. But twins belonging to different sets (and so having different gene combinations) are often very different.

IDENTICAL TWINS IN MAN

Identical twins are produced in man by the division of a single egg into two, so that they may be called monozygotic or 'one-egg' twins. There occur also dizygotic or 'two-egg' twins, produced from two separately fertilized eggs; these are often called fraternal twins. Methods have been discovered by which it is possible to determine to which class any pair of twins belongs.

Identical twins in man are of enormous interest; they supply exactly what is required for discovering what characteristics depend on genes. They are two individuals that are made throughout from the same set of materials, from the same set of genes. Every single gene of one has its exact duplicate in the other. The two therefore have the same heredity; the differences between them are due to environment alone. They show how like or how diverse will be two human beings that do not differ in their genes, in their heredity, although they may differ in the conditions of their lives, in the experiences they undergo. Identical twins thus form one of the most interesting and illuminating of the phenomena that nature presents to us. They yield the key to that ancient question, the relative role of nature and nurture: of heredity and environment.

The existence of the two-egg twins, with their diverse sets of genes, for comparison with the one-egg twins, having identical sets of genes, completes what is required for discovering just what resemblances and differences in human beings are due to identity and diversity of genes. The conditions to which the two members of a pair of twins are subjected are as much alike in the two-egg twins as in the one-egg twins. Therefore we can be certain that any differences between the two types of twins are due to differences in the genes, not to diversities of environment.Comparison of one-egg twins (having the same genes), with two-egg twins (having different genes): Identical twins are extraordinarily alike in all physical respects. They are always of the same sex. Their faces and Figs are so much alike that it is difficult to distinguish one from the other. They are alike in eye programmes, hair programmes, skin programmes. Their facial features are of the same form and relative size. Their teeth are alike, even to the irregularities. Their stature and weight are very similar. Their finger prints are as much alike as those of the two hands of the same individual.In all these respects the two-egg twins are often very different. They are no more alike than any other two members of the same family.Identical twins are also alike in weaknesses, in tendencies to certain diseases. If one has tuberculosis or other constitutional disease at a certain time of life, the other one usually has it also at about the same time, even though the twins live apart. But two-egg twins do not show these resemblances.

All this demonstrates that all these physical characteristics in man depend on genes. They are alike in two individuals whose genes are alike, diverse in two individuals whose genes are diverse.One-egg twins at times show certain physical differences, resulting either from certain features of the process of splitting the original egg from which the two were produced, or from the fact that one of the twins has been injured by accident or disease. From the former cause arise the facts that (1) sometimes one of the twins is right-handed, the other left-handed, and (2) one is sometimes more vigorous than the other. If the splitting of the original single egg occurs very early, before the right and left sides have become diverse, the two twins are alike in 'handedness' and in vigour; if the splitting occurs later they may be diverse in these respects. The effects of diversity of environment on twins we take up in a later chapter.

MENTAL CHARACTERISTICS IN TWINS

In a number of cases the two twins from a single egg have been separated while very young ýÿbefore the age of one or two years, and have lived apart Afterwards. In such cases the genes of the two individuals are alike, while their environments have been different. What is the situation as to their mental characteristics?

Several cases of this sort have been very carefully studied by Muller and by Newman. We shall have occasion to look at the results of this study in some detail later. Here we may summarize briefly from our present point of view what was discovered. One-egg twins that have lived apart show the same close resemblance in physical characterization that is found in such twins as have lived together. They show also resemblance in mental characteristics: in emotions, temperament, and in 'intelligence' as measured by psychological tests. In certain cases, there are found some dissimilarities in mental and emotional characteristics, depending on differences in education and experience in the separated twins.

Thus it is clear that (1) the genes have great influence on mental characteristics, emotions, temperament, and the like, but that (2) these things may be modified by different education and experience.

An illuminating study of the mental and moral characteristics of twins was made by Johannes Lange. He found in the prisons of Germany, or in the prison records, thirty persons, each of whom was a member of a pair of twins, the two members in each case being of the same sex. He traced and examined the other twin of each pair. Thus one twin was chosen because he was a criminal; the other twin was taken as he chanced to be. As it turned out, thirteen of the pairs were single-egg twins, while the other seventeen were two-egg twins.

Of the thirteen pairs of one-egg twins, one of which was known in each case to be a criminal, it turned out, upon investigation, that the other also had a criminal record in 10 out of the 13 cases. It is clear that if one such twin comes in conflict with the law, the other one, identical in original nature, has small chance of escaping that fate. But with the two-egg twins, made up of diverse original materials, the situation was different. Their environments were as similar as were those of the one-egg twins. But of the 17 pairs of this kind, in only three cases was the other member of the pair found to have a criminal record. In the other 14 pairs the fate of the two was quite diverse. One was a criminal in each case; the other was not.

Thus it is clear that everything that helps to determine whether a man shall be a criminal or a good citizen is deeply affected by the materials of which the individual is made, by his 'heredity'. Mentality, morality, conduct, career, fate, are affected by the hereditary materials with which the individual starts life. The truth of this conclusion came out still more strongly when the detailed records of the twins were studied. In the case of the single-egg twins, the two members conducted themselves as duplicate personalities might be expected to do. One such twin was a burglar; on looking up his brother's record, he was found to be a burglar too. In another pair, one twin was a high-powered financial swindler, fraudulently collecting large sums from his dupes. And so also was the other twin of that pair. Besides the pair that were burglars, and the pair that were swindlers, another pair of one-egg twins both 'committed puerile offences against the property laws'. Another consisted, in the words of Lange, of 'guttersnipes, but good fellows at heart', though they 'cannot stand alcohol; it develops frenzy and draws the knives from their pockets'. Another pair described 'have too little sense and will power', while in still another, both 'are lacking in all human feeling except for their own unpleasant selves'. And so goes the account throughout the entire list; the similarity in conduct is astonishing. Lange says, 'In all these pairs the type of crime is identical, the criminal careers begin at about the same age, and the behaviour of both members in court and in prison corresponds absolutely.'

But in the two-egg twins, that started life with different parental materials, with different heredity, the situation was different. As we have seen, 14 out of the 17 had totally different careers, one being a criminal, the other not. And there was no such detailed parallelism in the records of the two as was found in the one-egg twins. As before remarked, the similarity of environment was as great in the two-egg twins as it was in those that came from one egg. But this similarity of environment did not result in identity of fate or in detailed parallelism in careers. It was the identity of original nature, of the materials out of which they are made, that gave the same careers and the same fate to the one-egg twins. They were almost as much alike in their behaviour, their mentality, their career and their fate, as they were in their physical featuresýÿin which, like other one-egg twins, they were almost indistinguishable.

It is certain, therefore, that all the things that affect character and conduct are deeply influenced by the hereditary materials. There is no characteristic or quality of human beings that is exempted from its influence. This conclusion is confirmed by all the many studies that have been made on the two types of twins. And it is in harmony too with all that we know of the science of genetics. Experimental work on the breeding of other organisms shows, as we have seen, that they have no characteristics of any kind that are not affected by the genes that they receive at the beginning of their lives.

Are characteristics influenced also by other conditions, by the environment? The fact that all are affected by genes does not at all imply that they may not be affected in other ways.

5

Genes Heredity

HEREDITY

HISTORICAL PERSPECTIVE

For much of human history people were unaware of the scientific details of how babies were conceived and how heredity worked. Clearly they were conceived, and clearly there was some hereditary connection between parents and children, but the mechanisms were not readily apparent. The Greek philosophers had a variety of ideas: Theophrastus proposed that male flowers caused female flowers to ripen; Hippocrates speculated that "seeds" were produced by various body parts and transmitted to offspring at the time of conception, and Aristotle thought that male and female semen mixed at conception. Aeschylus, in 458 BC, proposed the male as the parent, with the female as a "nurse for the young life sown within her".

Dutch microscopist Anton van Leeuwenhoek discovered "animalcules" in the sperm of humans and other animals. Some scientists speculated they saw a "little man" (homunculus) inside each sperm. These scientists formed a school of thought known as the "spermists". They contended the only contributions of the female to the next generation were the womb in which the homunculus grew, and prenatal influences of the womb.

An opposing school of thought, the ovists, believed that the future human was in the egg, and that sperm merely stimulated the growth of the egg. Ovists thought women carried eggs containing boy and girl children, and that the gender of the offspring was determined well before conception. Pangenesis was an idea that males and females formed "pangenes" in every organ. These pangenes subsequently moved through their blood to the genitals and then to the children. The concept originated with the ancient Greeks and influenced biology until little over 100 years ago. The terms "blood relative", "full-blooded", and "royal blood" are relicts of pangenesis. Francis Galton, Charles Darwin's cousin, experimentally tested and disproved pangenesis during the 1870s. Blending theories of inheritance supplanted the spermists and ovists during the 19th century. The mixture of sperm and egg resulted in progeny that were a "blend" of two parents' characteristics. Sex cells are known collectively as gametes (*gamos*, Greek, meaning marriage). According to the blenders, when a black furred animal mates with white furred animal, you would expect all resulting progeny would be gray (a colour intermediate between black and white). This is often not the case.

Blending theories ignore characteristics skipping a generation. Charles Darwin had to deal with the implications of blending in his theory of evolution. He was forced to recognize blending as not important (or at least not the major principle), and suggest that science of the mid-1800s had not yet got the correct answer. That answer came from a contemporary, Gregor Mendel, although Darwin apparently never knew of Mendel's work.

THE MONK AND HIS PEAS

An Austrian monk, Gregor Mendel, developed the fundamental principles that would become the modern science of genetics. Mendel demonstrated that heritable properties are parceled out in discrete units, independently inherited. These eventually were termed genes.

Mendel reasoned an organism for genetic experiments should have:

- A number of different traits that can be studied
- Plant should be self-fertilizing and have a flower structure that limits accidental contact
- Offspring of self-fertilized plants should be fully fertile.

Mendel's experimental organism was a common garden pea (*Pisum sativum*), which has a flower that lends itself to self-pollination. The male parts of the flower are termed the anthers.

They produce pollen, which contains the male gametes (sperm). The female parts of the flower are the stigma, style, and ovary. The egg (female gamete) is produced in the ovary. The process of pollination (the transfer of pollen from anther to stigma) occurs prior to the opening of the pea flower. The pollen grain grows a pollen tube which allows the sperm to travel through the stigma and style, eventually reaching the ovary.

The ripened ovary wall becomes the fruit (in this case the pea pod). Most flowers allow cross-pollination, which can be difficult to deal with in genetic studies if the male parent plant is not known. Since pea plants are self-pollinators, the genetics of the parent can be more easily understood. Peas are also self-compatible, allowing self-fertilized embryos to develop as readily as out-fertilized embryos.

Mendel tested all 34 varieties of peas available to him through seed dealers. The garden peas were planted and studied for eight years. Each character studied had two distinct forms, such as tall or short plant height, or smooth or wrinkled seeds. Mendel's experiments used some 28,000 pea plants.

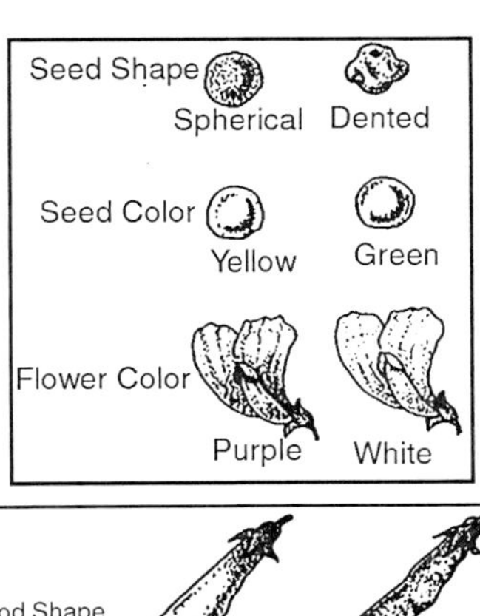

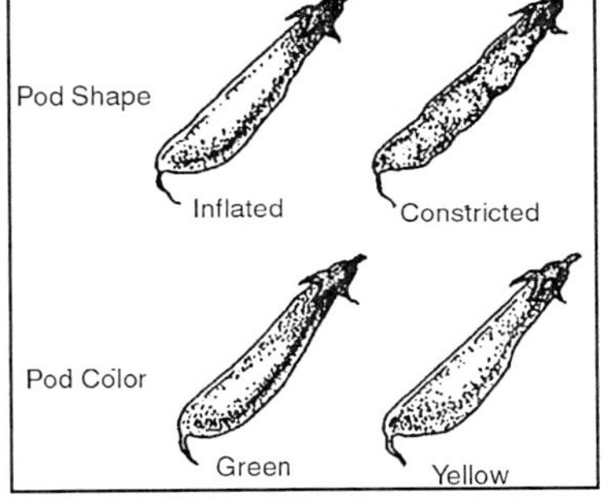

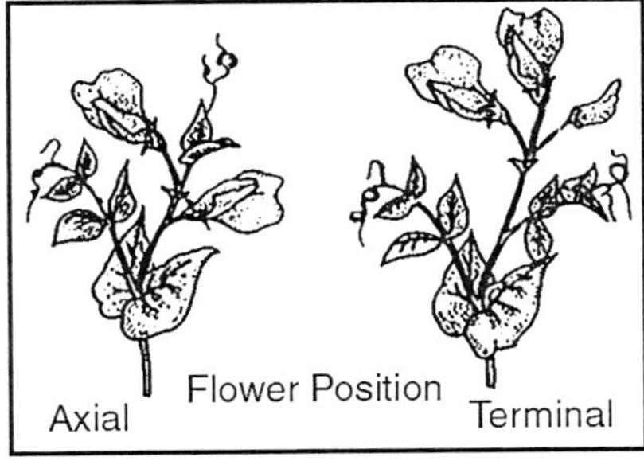

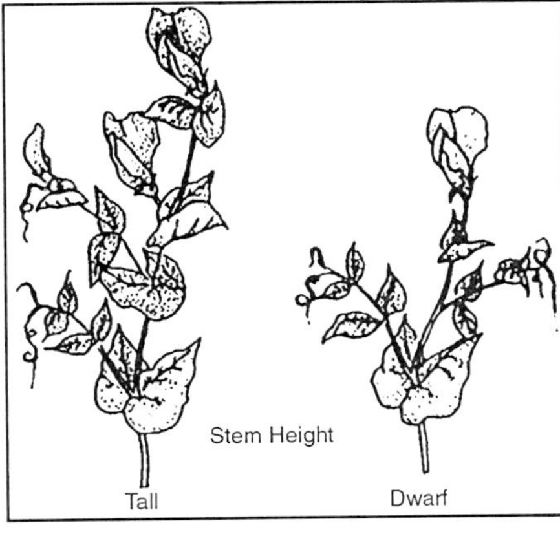

Fig. Some of Mendel's Traits as Expressed in Garden Peas.

Mendel's contribution was unique because of his methodical approach to a definite problem, use of clear-cut variables and application of mathematics (statistics) to the problem. Gregor Using pea plants and statistical methods, Mendel was able to demonstrate that traits were passed from each parent to their offspring through the inheritance of genes.

Mendel's work showed:

- Each parent contributes one factor of each trait shown in offspring.
- The two members of each pair of factors segregate from each other during gamete formation.
- The blending theory of inheritance was discounted.
- Males and females contribute equally to the traits in their offspring.
- Acquired traits are not inherited.

PRINCIPLE OF SEGREGATION

Mendel studied the inheritance of seed shape first. A cross involving only one trait is referred to as a monohybrid cross. Mendel crossed pure-breeding (also referred to as true-breeding) smooth-seeded plants with a variety that had always produced wrinkled seeds (60 fertilizations on 15 plants). All resulting seeds were smooth.The following year, Mendel planted these seeds and allowed them to self-fertilize. He recovered 7324 seeds: 5474 smooth and 1850 wrinkled. To help with record keeping, generations were labeled and numbered. The parental generation is denoted as the P1 generation.

The offspring of the P1 generation are the F_1 generation (first filial). The self-fertilizing F_1 generation produced the F_2 generation (second filial).

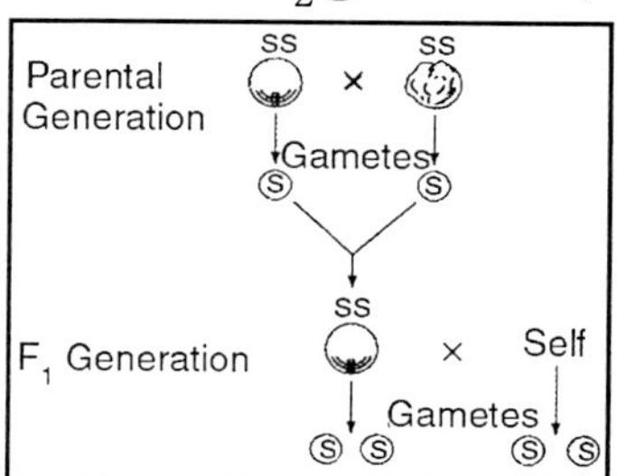

Inheritance of two alleles, S and s, in peas. Image from Purves et al., Life:

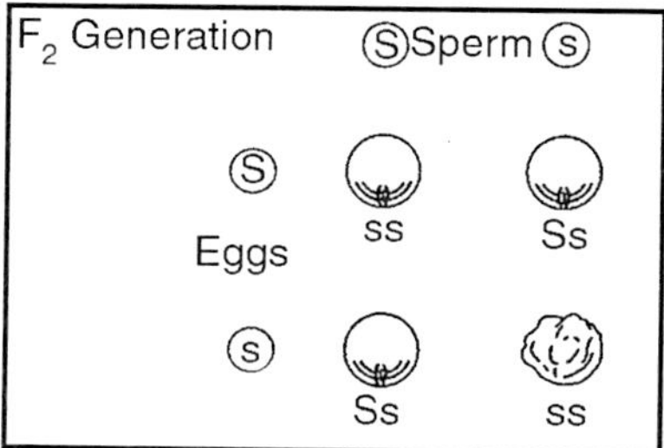

Fig. Punnett Square Explaining the Behaviour of the S and s Alleles.

P_1: smooth X wrinkled
F_1: all smooth
F_2: 5474 smooth and 1850 wrinkled

Meiosis, a process unknown in Mendel's day, explains how the traits are inherited.

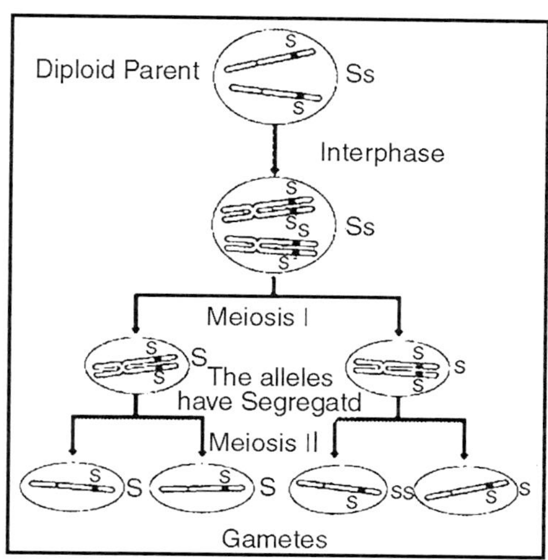

Fig. The Inheritance of the S and s Alleles Explained in light of Meiosis. Image from Purves et al., Life:

Mendel studied seven traits which appeared in two discrete forms, rather than continuous characters which are often difficult to distinguish. When "true-breeding" tall plants were crossed with "true-breeding" short plants, all of the offspring were tall plants.

The parents in the cross were the P1 generation, and the offspring represented the F_1 generation. The trait referred to as tall was considered dominant, while short was recessive. Dominant traits were defined by Mendel as those which appeared in the F1 generation in crosses between true-breeding strains. Recessives were those which "skipped" a generation, being expressed only when the dominant trait is absent. Mendel's plants exhibited complete dominance, in which the phenotypic expression of alleles was either dominant or recessive, not "in between".

When members of the F_1 generation were crossed, Mendel recovered mostly tall offspring, with some short ones also occurring. Upon statistically analyzing the F_2 generation, Mendel determined the ratio of tall to short plants was approximately 3:1. Short plants have skipped the F_1 generation, and show up in the F_2 and succeeding generations. Mendel concluded that the traits under study were governed by discrete (separable) factors. The factors were inherited in pairs, with each generation having a pair of trait factors. We now refer to these trait factors as alleles. Having traits inherited in pairs allows for the observed phenomena of traits "skipping" generations.

Summary of Mendel's Results:

- The F1 offspring showed only one of the two parental traits, and always the same trait.
- Results were always the same regardless of which parent donated the pollen (was male).
- The trait not shown in the F_1 reappeared in the F_2 in about 25per cent of the offspring.
- Traits remained unchanged when passed to offspring: they did not blend in any offspring but behaved as separate units.
- Reciprocal crosses showed each parent made an equal contribution to the offspring.

Mendel's Conclusions:

- Evidence indicated factors could be hidden or unexpressed, these are the recessive traits.
- The term phenotype refers to the outward appearance of a trait, while the term genotype is used for the genetic makeup of an organism.
- Male and female contributed equally to the offsprings' genetic makeup: therefore the number of traits was probably two (the simplest solution).
- Upper case letters are traditionally used to denote dominant traits, lower case letters for recessives.

Mendel reasoned that factors must segregate from each other during gamete formation (remember, meiosis was not yet known!) to retain the number of traits at 2. The Principle of Segregation proposes the separation of paired factors during gamete formation, with each gamete receiving one or the other factor, usually not both. Organisms carry two alleles for every trait. These traits separate during the formation of gametes.

DIHYBRID CROSSES

When Mendel considered two traits per cross (dihybrid, as opposed to single-trait-crosses, monohybrid), The resulting (F_2) generation did not have 3:1 dominant:recessive phenotype ratios. The two traits, if considered to inherit independently, fit into the principle of segregation. Instead of 4 possible genotypes from a monohybrid cross, dihybrid crosses have as many as 16 possible genotypes.

♀ gametes \ ♂ gametes	RY $\frac{1}{4}$	Ry $\frac{1}{4}$	ry $\frac{1}{4}$	rY $\frac{1}{4}$
RY $\frac{1}{4}$	RR YY $\frac{1}{16}$	RR Yy $\frac{1}{16}$	Rr Yy $\frac{1}{16}$	Rr YY $\frac{1}{16}$
Ry $\frac{1}{4}$	RR Yy $\frac{1}{16}$	RR yy $\frac{1}{16}$	Rr yy $\frac{1}{16}$	Rr Yy $\frac{1}{16}$
ry $\frac{1}{4}$	Rr Yy $\frac{1}{16}$	Rr yy $\frac{1}{16}$	rr yy $\frac{1}{16}$	rr Yy $\frac{1}{16}$
rY $\frac{1}{4}$	Rr YY $\frac{1}{16}$	Rr Yy $\frac{1}{16}$	rr Yy $\frac{1}{16}$	rr YY $\frac{1}{16}$

9 : 3 : 3 : 1

Round, yellow
Wrinkled, yellow
Round, green
Wrinkled, green

Mendel realized the need to conduct his experiments on more complex situations. He performed experiments tracking two seed traits: shape and colour. A cross concerning two traits is known as a dihybrid cross.

Crosses With Two Traits

Smooth seeds (S) are dominant over wrinkled (s) seeds.

Yellow seed colour (Y) is dominant over green (g).

Inheritance of two traits simultaneously, a dihybrid cross. The above graphic is from the Genetics.

Again, meiosis helps us understand the behaviour of alleles.

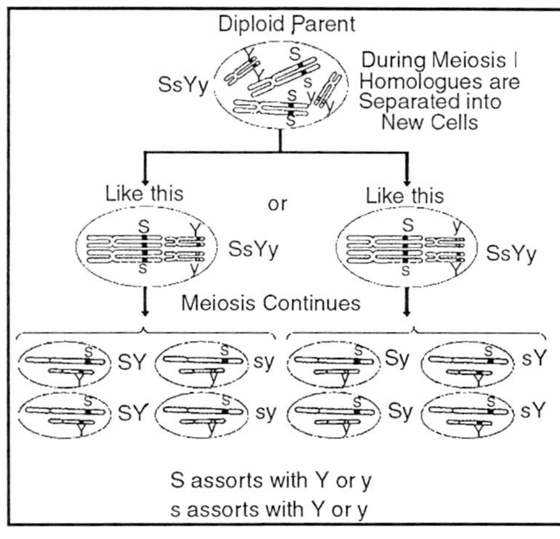

The inheritance of two traits on different chromosomes can be explained by meiosis. Image from Purves et al., Life:.

Methods, Results, and Conclusions

Mendel started with true-breeding plants that had smooth, yellow seeds and crossed them with true-breeding plants having green, wrinkled seeds. All seeds in the F_1 had smooth yellow seeds. The F_2 plants self-fertilized, and produced four phenotypes:

315 smooth yellow
108 smooth green
101 wrinkled yellow

32 wrinkled Green

Mendel analysed each trait for separate inheritance as if the other trait were not present.The 3:1 ratio was seen separately and was in accordance with the Principle of Segregation. The segregation of S and s alleles must have happened independently of the segregation of Y and y alleles. The chance of any gamete having a Y is 1/2; the chance of any one gamete having a S is 1/ 2.The chance of a gamete having both Y and S is the product of their individual chances (or 1/2 X 1/2 = 1/4). The chance of two gametes forming any given genotype is 1/4 X 1/4 (remember, the product of their individual chances). Thus, the Punnett Square has 16 boxes. Since there are more possible combinations to produce a smooth yellow phenotype (SSYY, SsYy, SsYY, and SSYy), that phenotype is more common in the F_2.

From the results of the second experiment, Mendel formulated the Principle of Independent Assortment — that when gametes are formed, alleles assort independently. If traits assort independent of each other during gamete formation, the results of the dihybrid cross can make sense. Since Mendel's time, scientists have discovered chromosomes and DNA. We now interpret the Principle of Independent Assortment as alleles of genes on different chromosomes are inherited independently during the formation of gametes. This was not known to Mendel. Punnett squares deal only with probability of a genotype showing up in the next generation. Usually if enough offspring are produced, Mendelian ratios will also be produced.

- Step 1 - definition of alleles and determination of dominance.
- Step 2 - determination of alleles present in all different types of gametes.
- Step 3 - construction of the square.
- Step 4 - recombination of alleles into each small square.
- Step 5 - Determination of Genotype and Phenotype ratios in the next generation.
- Step 6 - Labeling of generations, for example P_1, F_1, etc.

While answering genetics problems, there are certain forms and protocols that will make unintelligible problems easier to do. The term "true-breeding strain" is a code word for homozygous. Dominant alleles are those that show up in the next generation in crosses between two different "true-breeding strains". The key to any genetics problem is the recessive phenotype (more properly the phenotype that represents the recessive genotype). It is that organism whose genotype can be determined by examination of the phenotype. Usually homozygous dominant and heterozygous individuals have identical phenotypes (although their genotypes are different). This becomes even more important in dihybrid crosses.

MUTATIONS

Hugo de Vries, one of three turn-of-the-century scientists who rediscovered the work of Mendel, recognized that occasional abrupt, sudden changes occurred in the patterns of inheritance in the primrose plant. These sudden changes he termed mutations. De Vries proposed that new alleles arose by mutations. Charles Darwin, in his Origin of Species, was unable to describe how heritable changes were passed on to subsequent generations, or how new adaptations arose. Mutations provided answers to problems of the appearance of novel adaptations. The patterns of Mendelian inheritance explained the perseverance of rare traits in organisms, all of which increased variation, as you recall that was a major facet of Darwin's theory.

Mendel's work was published in 1866 but not recognized until the early 1900s when three scientists independently verified his principles, more than twenty years after his death. Ignored by the scientific community during his lifetime, Mendel's work is now a topic enjoyed by generations of biology students (;))

GENETIC TERMS

Definitions of terms. While we are discussing Mendel, we need to understand the context of his times as well as how his work fits into the modern science of genetics.

- Gene - a unit of inheritance that usually is directly responsible for one trait or character.
- Allele - an alternate form of a gene. Usually there are two alleles for every gene, sometimes as many a three or four.
- Homozygous - when the two alleles are the same.
- Heterozygous - when the two alleles are different, in such cases the dominant allele is expressed.
- Dominant - a term applied to the trait (allele) that is expressed irregardless of the second allele.
- Recessive - a term applied to a trait that is only expressed when the

second allele is the same (*e.g.* short plants are homozygous for the recessive allele).

- Phenotype - the physical expression of the allelic composition for the trait under study.
- Genotype - the allelic composition of an organism.
- Punnett squares - probability diagram illustrating the possible offspring of a mating.

GENE INTERACTIONS

FINDING THE GENES

Between 1884 (the year Mendel died) and 1888 details of mitosis and meiosis were reported, the cell nucleus was identified as the location of the genetic material, and "qualities" were even proposed to be transmitted on chromosomes to daughter cells at mitosis. In 1903 Walter Sutton and Theodore Boveri formally proposed that chromosomes contain the genes. The Chromosome Theory of Inheritance is one of the foundations of genetics and explains the physical reality of Mendel's principles of inheritance.

Fig. Fruit Fly, *Drosophila Melanogaster.*

The location of many genes (Mendel's factors) was determined by Thomas Hunt Morgan and his coworkers in the early 1900's. Morgan's experimental organism was the fruit fly (*Drosophila melanogaster*). Fruit flies are ideal organisms for genetics, having a small size, ease of care, susceptibility to mutations, and short (7-9 day) generation time. The role of chromosomes in determination of sex was deduced by Morgan from work on fruit flies.

During Metaphase I, homologous chromosomes will line up. A karyotype can be made by cutting and arranging photomicrographs of the homologous chromosomes thus revealed at Metaphase I.

Two types of chromosome pairs occur. Autosomes resemble each other in size and placement of the centromere, for example pairs of chromosome 21 are the same size, while pairs of chromosome 9 are of a different size from pair 21. Sex chromosomes may differ in their size, depending on the species of the organism they are from. In humans and *Drosophila*, males have a smaller sex chromosome, termed the Y, and a larger one, termed the X.

Males are thus XY, and are termed heterogametic. Females are XX, and are termed homogametic. In grasshoppers, which Sutton studied in discovering chromosomes, there is no Y, only the X chromosome in males. Females are

XX, while males are denoted as XO. Other organisms (notably birds, moths and butterflies) have males homogametic and females heterogametic. Males (if heterogametic) contribute either an X or Y to the offspring, while females contribute either X. The male thus determines the sex of the offspring. Remember that in meiosis, each chromosome is replicated and one copy sent to each gamete.

Morgan discovered a mutant eye colour and attempted to use this mutant as a recessive to duplicate Mendel's results. He failed, instead of achieving a 3:1 F2 ratio the ratio was closer to 4:1 (red to white). Most mutations are usually recessive, thus the appearance of the white mutant presented Morgan a chance to test Mendel's ratios on animals

. The F1 generation also had no white eyed females. Morgan hypothesized that the gene for eye colour was only on the X chromosome, specifically in that region of the X that had no corresponding region on the Y.

White eyed fruit flies were also more likely to die prior to adulthood, thus explaining the altered ratios. Normally eyes are red, but a variant (white) eyed was detected and used in genetic study. Cross a homozygous white eyed male with a homozygous red eyed female, and all the offspring have red eyes.

Red is dominant over white. However, cross a homozygous white eyed female with a red eyed male, and the unexpected results show all the males have white eyes and all the females red eyes. This can be explained if the eye colour gene is on the X chromosome.

Explanation

If the gene for eye colour is on the X chromosome, the red eyed male in the second cross will pass his red eyed X to only his daughters, who in turn received only a recessive white-carrying X from their mother. Thus all females had red eyes like their father. Since the male fruit fly passes only the Y to his sons, their eye colour is determined entirely by the single X chromosome they receive from their mother (in this case white). Thus all the males in the second cross were white eyed.

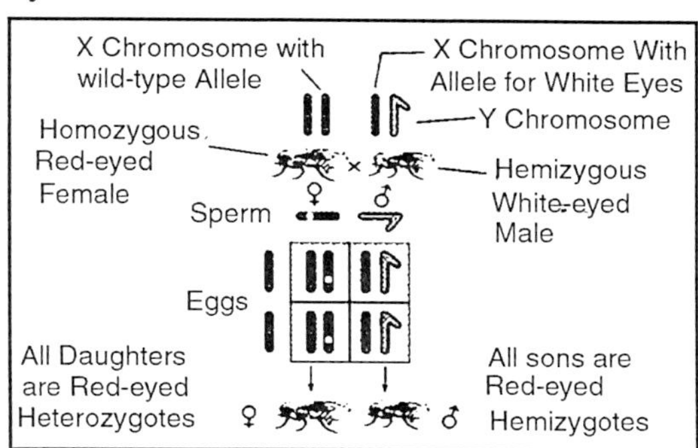

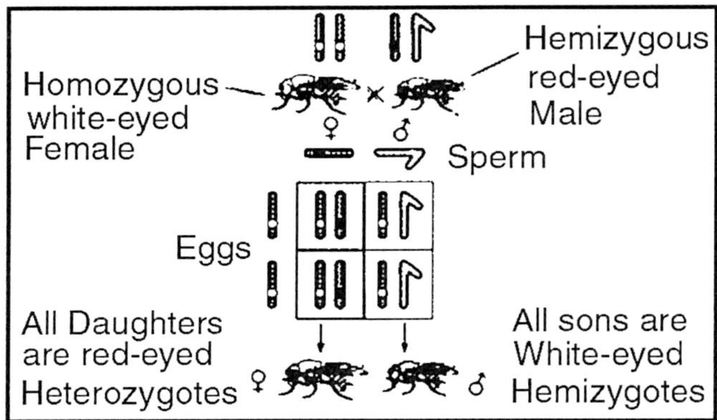

Fig. Inheritance of Eye Colour in Fruit flies.

These experiments introduced the concept of sex-linkage, the occurrence of genes on that part of the X that lack a corresponding location on the Y. Sex-linked recessives (such as white eyes in fruit flies, hemophilia, baldness, and colourblindness in humans) occur more commonly in males, since there is no chance of them being heterozygous. Such a condition is termed homozygous.

Characteristics of X-linked Traits

- Phenotypic expression more common in males
- Sons cannot inherit the trait from their fathers, but daughters can.

Sons inherit their Y chromosome from their father.

Only a few genes have been identified on the Y chromosome, among them the testis-determining factor (TDF) that promotes development of the male phenotype.

Barr bodies are interpreted as inactivated X chromosomes in mammalian females. Since females have two X chromosomes, the Lyon hypothesis suggests that one or the other X is inactivated in each somatic (non-reproductive) cell during embryonic development. Cells mitotically produced from these embryonic cells likewise have the same inactivated X chromosome.

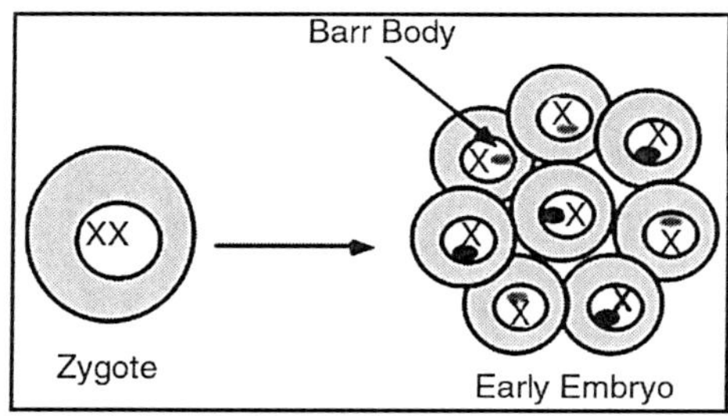

Fig. The Role of Deactivated X Chromosomes in Mammalian Female Development.).

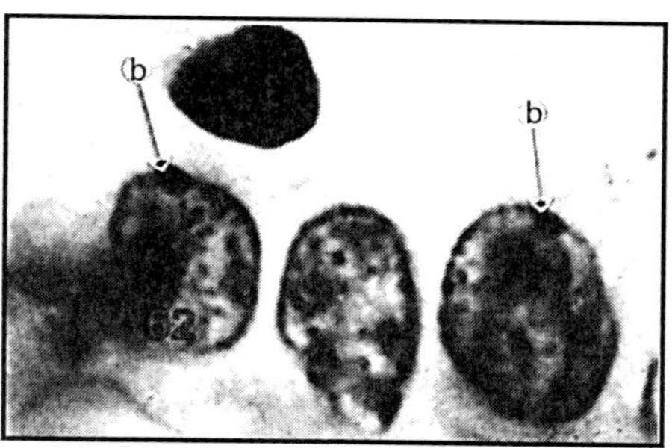

Fig. Barr Bodies (b) in Mammalian Cells.

Calico cats (sometimes called tortoiseshell) are almost always female since the calico trait is caused by some areas of the cat's fur expressing one allele and others expressing the other colour. Can there be a male tortoiseshell cat? How would such a cat get its genes? Remember that fur colour in cats is a sex-linked feature. Would the male calico be fertile or sterile?

THE MODERN VIEW OF THE GENE

While Mendel discussed traits, we now know that genes are segments of the DNA that code for specific proteins. These proteins are responsible for the expression of the phenotype. The basic principles of segregation and independent assortment as worked out by Mendel are applicable even for sex-linked traits.

Codominant alleles

Codominant alleles occur when rather than expressing an intermediate phenotype, the heterozygotes express both homozygous phenotypes. An example is in human ABO blood types, the heterozygote AB type manufactures antibodies to both A and B types. Blood Type A people manufacture only anti-B antibodies, while type B people make only anti-A antibodies. Codominant alleles are both expressed. Heterozygotes for codominant alleles fully express both alleles. Blood type AB individuals produce both A and B antigens. Since neither A nor B is dominant over the other and they are both dominant over O they are said to be codominant.

Incomplete Dominance

Incomplete dominance is a condition when neither allele is dominant over the other. The condition is recognized by the heterozygotes expressing an intermediate phenotype relative to the parental phenotypes. If a red flowered plant is crossed with a white flowered one, the progeny will all be pink. When pink is crossed with pink, the progeny are 1 red, 2 pink, and 1 white.

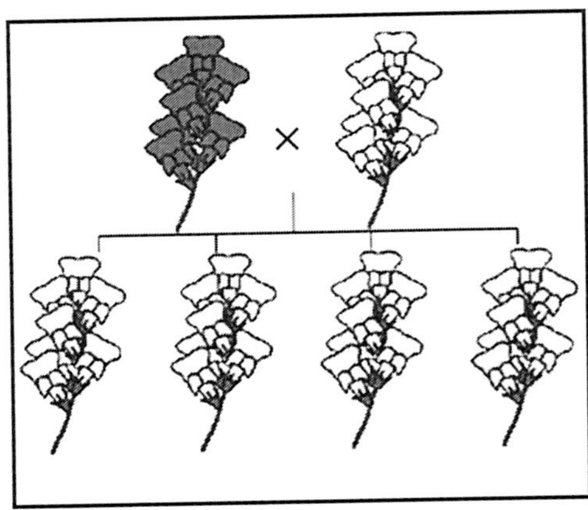

Fig. Inheritance of Flower Colour in Snapdragons

Flower colour in snapdragons is an example of this pattern. Cross a true-breeding red strain with a true-breeding white strain and the F_1 are all pink (heterozygotes). Self-fertilize the F1 and you get an F_2 ratio of 1 red: 2 pink: 1 white. This would not happen if true blending had occurred (blending cannot explain traits such as red or white skipping a generation and pink flowers crossed with pink flowers should produce ONLY pink flowers).

Multiple Alleles

Many genes have more than two alleles (even though any one diploid individual can only have at most two alleles for any gene), such as the ABO blood groups in humans, which are an example of multiple alleles. Multiple alleles result from different mutations of the same gene. Coat colour in rabbits is determined by four alleles. Human ABO blood types are determined by alleles A, B, and O. A and B are codominants which are both dominant over O. The only possible genotype for a type O person is OO. Type A people have either AA or AO genotypes. Type B people have either BB or BO genotypes. Type AB have only the AB (heterozygous) genotype.

The A and B alleles of gene I produce slightly different glycoproteins (antigens) that are on the surface of each cell. Homozygous A individuals have only the A antigen, homozygous B individuals have only the B antigen, homozygous O individuals produce neither antigen, while a fourth phenotype (AB) produces both A and B antigens.

INTERACTIONS AMONG GENES

While one gene may make only one protein, the effects of those proteins usually interact (for example widow's peak may be masked by expression of the baldness gene). Novel phenotypes often result from the interactions of two genes, as in the case of the comb in chickens. The single comb is produced only by the rrpp genotype. Rose comb (b) results from R_pp. (_ can be either R or r). Pea comb (c) results from rrP_. Walnut comb, a novel phenotype, is produced when the genotype has at least one dominant of each gene (R_P_).

Epistasis

Epitasis is the term applied when one gene interferes with the expression of another (as in the baldness/widow's peak mentioned earlier). Bateson reported a different phenotypic ratio in sweet pea than could be explained by simple Mendelian inheritance. This ratio is 9:7 instead of the 9:3:3:1 one would expect of a dihybrid cross between heterozygotes. Of the two genes (C and P), when either is homozygous recessive (cc or pp) that gene is epistatic to (or hides) the other. To get purple flowers one must have both C and P alleles present.

ENVIRONMENT AND GENE EXPRESSION

Phenotypes are always affected by their environment. In buttercup (*Ranunculus peltatus*), leaves below water-level are finely divided and those above water-level are broad, floating, photosynthetic leaf-like leaves. Siamese cats are darker on their extremities, due to temperature effects on phenotypic expression.

Expression of phenotype is a result of interaction between genes and environment. Siamese cats and Himalayan rabbits both animals have dark coloured fur on their extremities. This is caused by an allele that controls pigment production being able only to function at the lower temperatures of those extremities. Environment determines the phenotypic pattern of expression.

POLYGENIC INHERITANCE

Polygenic inheritance is a pattern responsible for many features that seem simple on the surface. Many traits such as height, shape, weight, colour, and metabolic rate are governed by the cumulative effects of many genes. Polygenic traits are not expressed as absolute or discrete characters, as was the case with Mendel's pea plant traits. Instead, polygenic traits are recognizable by their expression as a gradation of small differences (a continuous variation). The results form a bell shaped curve, with a mean value and extremes in either direction.Height in humans is a polygenic trait, as is colour in wheat kernels. Height in humans is NOT discontinuous. If you line up the entire class a continuum of variation is evident, with an average height and extremes in variation (very short [vertically challenged?] and very tall [vertically enhanced]). Traits showing continuous variation are usually controlled by the additive effects of two or more separate gene pairs. This is an example of polygenic inheritance. The inheritance of EACH gene follows Mendelian rules.

Usually polygenic traits are distinguished by:

- Traits are usually quantified by measurement rather than counting.

- Two or more gene pairs contribute to the phenotype.
- Phenotypic expression of polygenic traits varies over a wide range.

Human polygenic traits include:

- Height
- SLE (Lupus)
- Weight
- Eye Colour
- Intelligence
- Skin Colour
- Many forms of behaviour

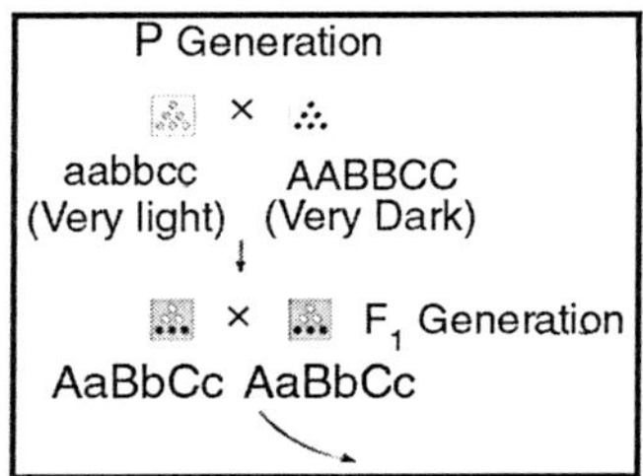

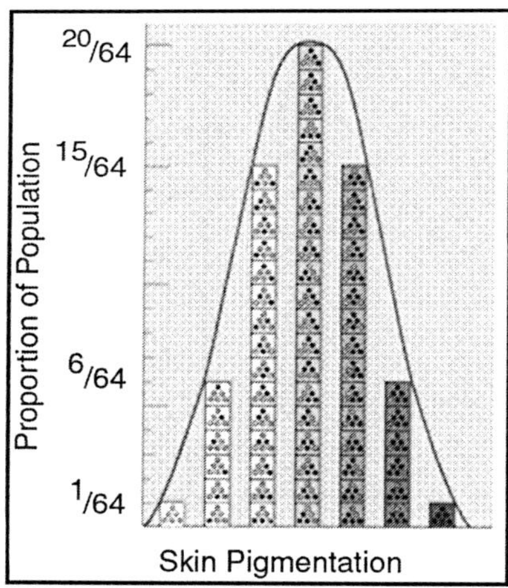

The role of many genes (polygenic inheritance) in production of a continuum of phenotypes. Images.

PLEIOTROPY

Pleiotropy is the effect of a single gene on more than one characteristic. An example is the "frizzle-trait" in chickens. The primary result of this gene is the production of defective feathers. Secondary results are both good and bad; good include increased adaptation to warm temperatures, bad include increased metabolic rate, decreased egg-laying, changes in heart, kidney and spleen. Cats that are white with blue eyes are often deaf, white cats with a blue and an yellow-orange eye are deaf on the side with the blue eye. Sickle-cell anemia is a human disease originating in warm lowland tropical areas where malaria is common. Sickle-celled individuals suffer from a number of problems, all of which are pleiotropic effects of the sickle-cell allele.

GENES AND CHROMOSOMES

Linkage occurs when genes are on the same chromosome. Remember that sex-linked genes are on the X chromosome, one of the sex chromosomes. Linkage groups are invariably the same number as the pairs of homologous chromosomes an organism possesses.Recombination occurs when crossing-over has broken linkage groups, as in the case of the genes for wing size and body colour that Morgan studied. Chromosome mapping was originally based on the frequencies of recombination between alleles.

Since mutations can be induced (by radiation or chemicals), Morgan and his coworkers were able to cause new alleles to form by subjecting fruit flies to mutagens (agents of mutation, or mutation generators). Genes are located on specific regions of a certain chromosome, termed the gene locus (plural: loci). A gene therefore is a specific segment of the DNA molecule.

Alfred Sturtevant, while an undergraduate student in Morgan's lab, postulated that crossing-over would be less common between genes adjacent to each other on the same chromosome and that it should be possible to plot the sequence of genes along a fruit fly chromosome by using crossing-over frequencies. Distances on gene maps are expressed in map units (one map unit = 1 recombinant per 100 fertilized eggs; or a 1per cent chance of recombination).

The map for *Drosophila melanogaster* chromosomes is well known. Note that eye colour and aristae length are far apart, as indicated by the occurrence of more recombinants (crossing-overs) between them, while wing length is closer to eye shape (as indicated by the low frequency of recombination between these two features).

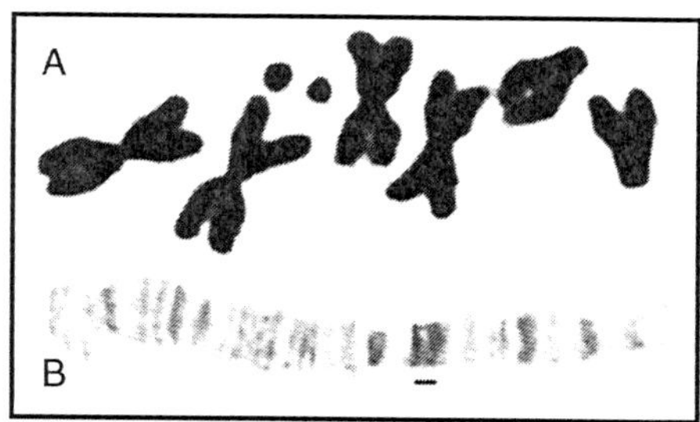

Fig. A Illustrates the Fruit fly Chromosomes at Metaphase; figure B shows a Polytene Chromosome.

CHROMOSOME ABNORMALITIES

Chromosome abnormalities include inversion, insertion, duplication, and deletion. These are types of mutations. Since DNA is information, and information typically has a beginning point, an inversion would produce an inactive or altered protein. Likewise deletion or duplication will alter the gene product.

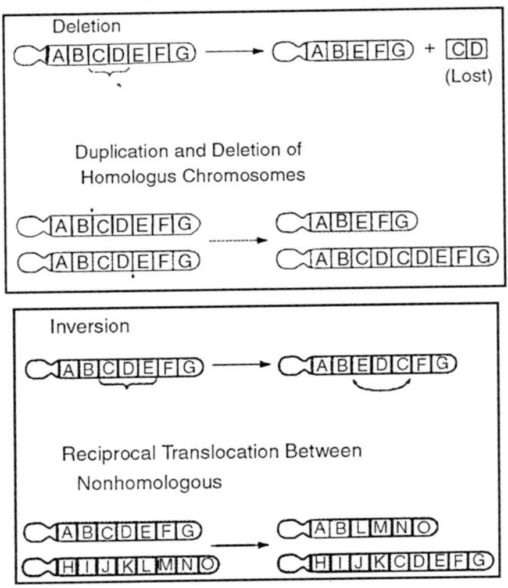

PRINCIPLES OF GENETICS

MENDEL'S GENETICS

For thousands of years farmers and herders have been selectively breeding their plants and animals to produce more useful hybrids. It was somewhat of a hit or miss process since the actual mechanisms governing inheritance were unknown. Knowledge of these genetic mechanisms finally came as a result of careful labouratory breeding experiments carried out over the last century and a half.

By the 1890's, the invention of better microscopes allowed biologists to discover the basic facts of cell division and sexual reproduction. The focus of genetics research then shifted to understanding what really happens in the transmission of hereditary traits from parents to children. A number of hypotheses were suggested to explain heredity, but Gregor Mendel , a little known Central European monk, was the only one who got it more or less right. His ideas had been published in 1866 but largely went unrecognized until 1900, which was long after his death. His early adult life was spent in relative obscurity doing basic genetics research and teaching high school mathematics, physics, and Greek in Brno (now in the Czech Republic). In his later years, he became the abbot of his monastery and put aside his scientific work.

While Mendel's research was with plants, the basic underlying principles of heredity that he discovered also apply to people and other animals because the mechanisms of heredity are essentially the same for all complex life forms.

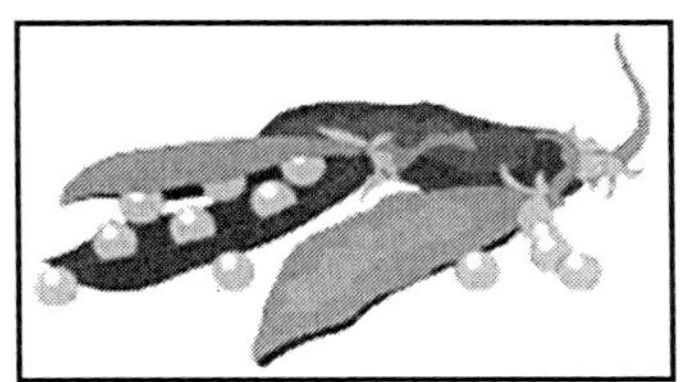

Fig. Common Edible Peas

Through the selective cross-breeding of common pea plants (*Pisum sativum*) over many generations, Mendel discovered that certain traits show up in offspring without any blending of parent characteristics. For instance, the pea flowers are either purple or white—intermediate colours do not appear in the offspring of cross-pollinated pea plants.

Mendel observed seven traits that are easily recognized and apparently only occur in one of two forms:

- Flower colour is purple or white
- Flower position is axil or terminal
- Stem length is long or short
- Seed shape is round or wrinkled
- Seed colour is yellow or green
- Pod shape is inflated or constricted
- Pod colour is yellow or green

This observation that these traits do not show up in offspring plants with intermediate forms was critically important because the leading theory in biology at the time was that inherited traits blend from generation to generation. Most of the leading scientists in the 19th century accepted this "blending theory."

Charles Darwin proposed another equally wrong theory known as "pangenesis". This held that hereditary "particles" in our bodies are affected by the things we do during our lifetime.These modified particles were thought to migrate via blood to the reproductive cells and subsequently could be inherited by the next generation. This was essentially a variation of Lamarck's incorrect idea of the "inheritance of acquired characteristics."Mendel picked common garden pea plants for the focus of his research because they can be grown easily in large numbers and their reproduction can be manipulated. Pea plants have both male and female reproductive organs. As a result, they can either self-pollinate themselves or cross-pollinate with another plant.

In his experiments, Mendel was able to selectively cross-pollinate purebred plants with particular traits and observe the outcome over many generations. This was the basis for his conclusions about the nature of genetic inheritance.

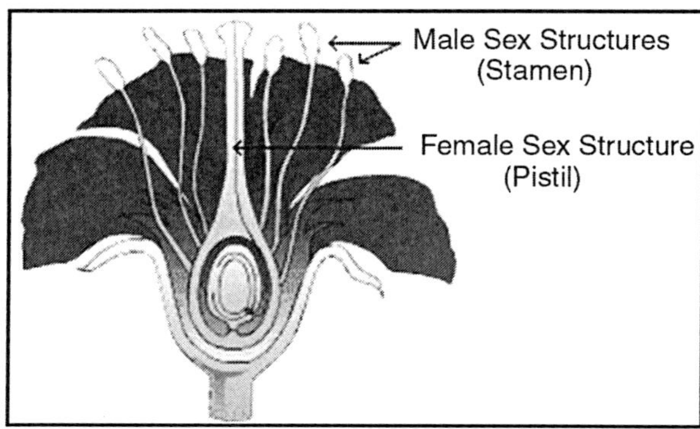

Fig. Reproductive Structures of Flowers

In cross-pollinating plants that either produce yellow or green pea seeds exclusively, Mendel found that the first offspring generation (f_1) always has yellow seeds. However, the following generation (f_2) consistently has a 3:1 ratio of yellow to green.

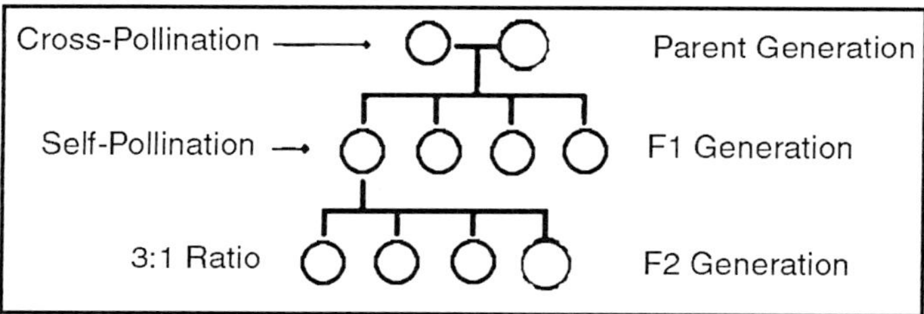

This 3:1 ratio occurs in later generations as well. Mendel realized that this was the key to understanding the basic mechanisms of inheritance.

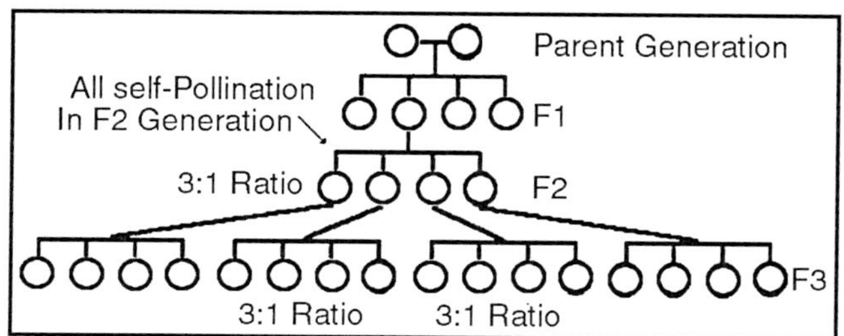

- The came to three important conclusions from these experimental results:
- That the inheritance of each trait is determined by "units" or "factors" that are passed on to descendents unchanged (these units are now called genes)
- That an individual inherits one such unit from each parent for each trait
- That a trait may not show up in an individual but can still be passed on to the next generation.

It is important to realize that, in this experiment, the starting parent plants were homozygous for pea seed colour. That is to say, they each had two identical forms (or alleles) of the gene for this trait—2 yellows or 2 greens. The plants in the f1 generation were all heterozygous. In other words, they each had inherited two different alleles—one from each parent plant. It becomes clearer when we look at the actual genetic makeup, or genotype, of the pea plants instead of only the phenotype, or observable physical characteristics.

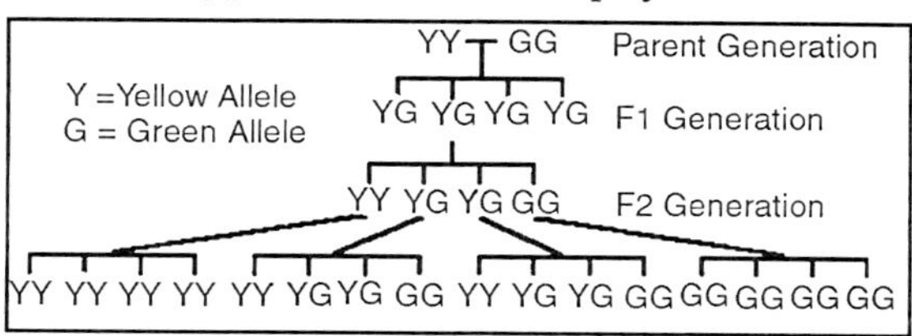

Note that each of the f1 generation plants inherited a Y allele from one parent and a G allele from the other. When the f1 plants breed, each has an equal chance of passing on either Y or G alleles to each offspring. With all of the seven pea plant traits that Mendel examined, one form appeared dominant over the other.Which is to say, it masked the presence of the other allele. For example, when the genotype for pea seed colour is YG (heterozygous), the phenotype is yellow. However, the dominant yellow allele does not alter the recessive green one in any way. Both alleles can be passed on to the next generation unchanged.

Mendel's observations from these experiments can be summarized in two principles:

1. The principle of segregation
2. The principle of independent assortment

According to the principle of segregation, for any particular trait, the pair of alleles of each parent separate and only one allele passes from each parent on to an offspring. Which allele in a parent's pair of alleles is inherited is a matter of chance. We now know that this segregation of alleles occurs during the process of sex cell formation (*i.e.*, meiosis).

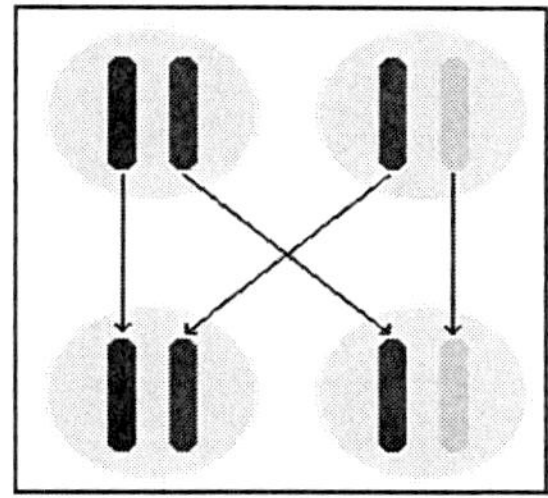

Fig. Segregation of Alleles in the Production of Sex Cells

According to the principle of independent assortment, different pairs of alleles are passed to offspring independently of each other. The result is that new combinations of genes present in neither parent are possible. For example, a pea plant's inheritance of the ability to produce purple flowers instead of white ones does not make it more likely that it will also inherit the ability to produce yellow pea seeds in contrast to green ones. Likewise, the principle of independent assortment explains why the human inheritance of a particular eye colour does not increase or decrease the likelihood of having 6 fingers on each hand. Today, we know this is due to the fact that the genes for independently assorted traits are located on different chromosomes.

PROBABILITY OF INHERITANCE

The value of studying genetics is in understanding how we can predict the likelihood of inheriting particular traits. This can help plant and animal breeders in developing varieties that have more desirable qualities. It can also help people explain and predict patterns of inheritance in family lines. One of the easiest ways to calculate the mathematical probability of inheriting a specific trait was invented by an early 20th century English geneticist named Reginald Punnett. His technique employs what we now call a Punnett square. This is a simple graphical way of discovering all of the potential combinations of genotypes that can occur in children, given the genotypes of their parents. It also shows us the odds of each of the offspring genotypes occurring.

Setting up and using a Punnett square is quite simple once you understand how it works. You begin by drawing a grid of perpendicular lines:

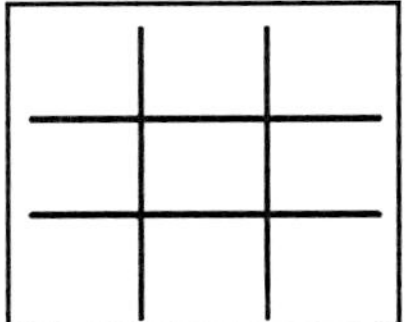

Next, you put the genotype of one parent across the top and that of the other parent down the left side. For example, if parent pea plant genotypes were YY and GG respectively, the setup would be:

Genetic Contribution of one Parent
G G
Genetic Contribution of the other Parent
Y
Y

Note that only one letter goes in each box for the parents. It does not matter which parent is on the side or the top of the Punnett square. Next, all you have to do is fill in the boxes by copying the row and column-head letters across or down into the empty squares. This gives us the predicted frequency of all of the potential genotypes among the offspring each time reproduction occurs.

Genetic Contribution of one Parent

	G	G
Y	YG	YG
Y	YG	YG

Genetic Contribution of the other Parent

Offspring Genotype Possibilities

In this example, 100per cent of the offspring will likely be heterozygous (YG). Since the Y (yellow) allele is dominant over the G (green) allele for pea plants, 100per cent of the YG offspring will have a yellow phenotype, as Mendel observed in his breeding experiments.

In another example (shown below), if the parent plants both have heterozygous (YG) genotypes, there will be 25per cent YY, 50per cent YG, and 25per cent GG offspring on average. These per centages are determined based on the fact that each of the 4 offspring boxes in a Punnett square is 25per cent (1 out of 4). As to phenotypes, 75per cent will be Y and only 25per cent will be G. These will be the odds every time a new offspring is conceived by parents with YG genotypes.

An offspring's genotype is the result of the combination of genes in the sex cells or gametes (sperm and ova) that came together in its conception. One sex cell came from each parent. Sex cells normally only have one copy of the gene for each trait (*e.g.*, one copy of the Y or G form of the gene in the example above). Each of the two Punnett square boxes in which the parent genes for a trait are placed (across the top or on the left side) actually represents one of the two possible genotypes for a parent sex cell. Which of the two parental copies of a gene is inherited depends on which sex cell is inherited--it is a matter of chance. By placing each of the two copies in its own box has the effect of giving it a 50per cent chance of being inherited.

If you are not yet clear about how to make a Punnett Square and interpret its result, take the time to try to figure it out before going on.

Are Punnett Squares Just Academic Games

Why is it important for you to know about Punnett squares? The answer is that they can be used as predictive tools when considering having children. Let us assume, for instance, that both you and your mate are carriers for a particularly unpleasant genetically inherited disease such as cystic fibrosis. Of course, you are worried about whether your children will be healthy and normal. For this example, let us define "A" as being the dominant normal allele and "a" as the recessive abnormal one that is responsible for cystic fibrosis. As carriers, you and your mate are both heterozygous (Aa). This disease only afflicts those who are homozygous recessive (aa). The Punnett square below makes it clear that at each birth, there will be a 25per cent chance of you having a normal homozygous (AA) child, a 50per cent chance of a healthy heterozygous (Aa) carrier child like you and your mate, and a 25per cent chance of a homozygous recessive (aa) child who probably will eventually die from this condition.

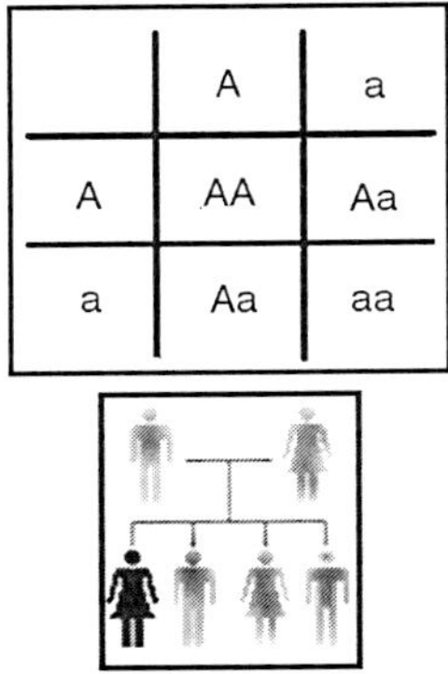

	A	a
A	AA	Aa
a	Aa	aa

If both parents are carriers of the recessive allele for a disorder, all of their children will face the following odds of inheriting it:

25per cent chance of having the recessive disorder

50per cent chance of being a healthy carrier

25per cent chance of being healthy and not have the recessive allele at all

If a carrier (Aa) for such a recessive disease mates with someone who has it (aa), the likelihood of their children also inheriting the condition is far greater (as shown below). On average, half of the children will be heterozygous (Aa) and, therefore, carriers. The remaining half will inherit 2 recessive alleles (aa) and develop the disease.

	A	a
a	Aa	aa
a	Aa	aa

If one parent is a carrier and the other has a recessive disorder, their children will have the following odds of inheriting it:

50per cent chance of being a healthy carrier

50per cent chance having the recessive disorder

It is likely that every one of us is a carrier for a large number of recessive alleles. Some of these alleles can cause life-threatening defects if they are inherited from both parents. In addition to cystic fibrosis, albinism, and beta-thalassemia are recessive disorders. Some disorders are caused by dominant alleles for genes. Inheriting just one copy of such a dominant allele will cause the disorder. This is the case with Huntington disease, achondroplastic dwarfism, and polydactyly. People who are heterozygous (Aa) are not healthy carriers. They have the disorder just like homozygous dominant (AA) individuals.

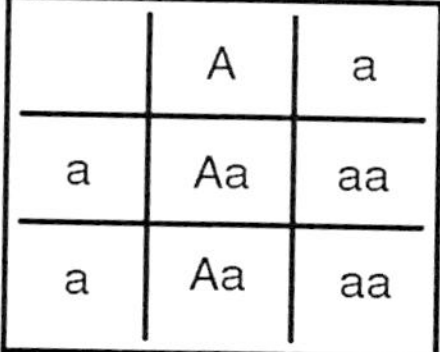

	A	a
a	Aa	aa
a	Aa	aa

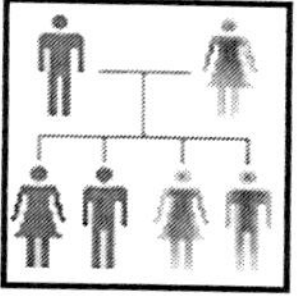

If only one parent has a single copy of a dominant allele for a dominant disorder, their children will have a 50per cent chance of inheriting the disorder and 50per cent chance of being entirely normal.

Punnett squares are standard tools used by genetic counselors. Theoretically, the likelihood of inheriting many traits, including useful ones, can be predicted using them. It is also possible to construct squares for more than one trait at a time. However, some traits are not inherited with the simple mathematical probability suggested here.

MODIFYING AND REGULATOR GENES

There are two classes of genes that can have an effect on how other genes function. They are called modifying genes and regulator genes.

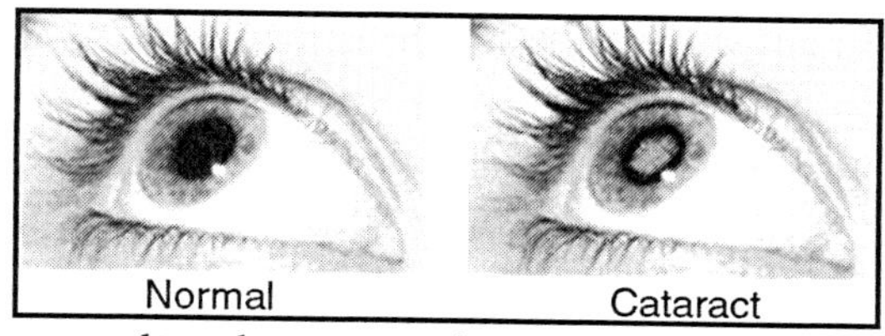

Modifying genes alter how certain other genes are expressed in the phenotype. For instance, there is a dominant cataract gene which will produce varying degrees of vision impairment depending on the presence of a specific allele for a companion modifying gene. However, cataracts also can be promoted by diabetes and common environmental factors such as excessive ultraviolet radiation, and alcoholism. Nearly half of all people in North America over 65 years of age eventually develop them.

Regulator genes can either initiate or block the expression of other genes. They control the production of a variety of chemicals in plants and animals. For instance, the time of production of specific proteins that will be new structural parts of our bodies can be controlled by such regulator genes. Shortly after conception, regulator genes work as master switches orchestrating the timely development of our body parts. They are also responsible for changes that occur in our bodies as we grow older. In other words, they control the maturation and aging processes. Regulator genes that are involved in subdividing an embryo into what will become the major body parts of an individual are also referred to as homeotic, homeobox, or Hox genes. They are responsible for setting generalized cells on the path to become a head, torso, arms, legs, etc.

Incomplete Penetrance

Some genes are incompletely penetrant. That is to say, their effect does not normally occur unless certain environmental factors are present. For example, you may inherit the genes that are responsible for type 2 diabetes but never get the disease unless you become greatly overweight, persistently stressed psychologically, or do not get enough sleep on a regular basis. Similarly, the genes that cause the chronic autoimmune disease, multiple sclerosis, may be triggered by the Epstein-Barr virus and possibly other specific environmental stresses. New research suggests that abundant exposure to the sun in childhood can provide some protection from developing MS. Subsequently, people who grow up in tropical and subtropical regions of the world have significantly lower rates of MS as adults.

Sex Related Genetic effects

There are three categories of genes that may have different effects depending on an individual's gender.

These are referred to as:

- Sex-limited genes
- Sex-controlled genes
- Genome imprinting

Sex-limited genes are ones that are inherited by both men and women but are normally only expressed in the phenotype of one of them. The heavy male beard is an example. While women have facial hair it is most often very fine and comparatively sparse.

In contrast, sex-controlled genes are expressed in both sexes but differently. An example of this is gout, a disease that causes painfully inflamed joints. If the gene is present, men are nearly eight times more likely than women to have severe symptoms. Some genes are known to have a different effect depending on the gender of the parent from whom they are inherited. This phenomenon is referred to as genome imprinting or genetic imprinting. Apparently, diabetes, psoriasis, and some rare genetically inherited diseases, such as a form of mental retardation known as Angelman syndrome, can follow this inheritance pattern.

Pleiotropy

A single gene may be responsible for a variety of traits. This is called pleiotropy. The complex of symptoms that are collectively referred to as sickle-cell trait, or sickle-cell anemia, is an example. A single gene results in irregularly shaped red blood cells that painfully block blood vessels, cause poor overall physical development, as well as related heart, lung, kidney, and eye problems. Another pleiotropic trait is albinism. The gene for this trait not only results in a deficiency of skin, hair, and eye pigmentation but also causes defects in vision.

Stuttering Alleles

Lastly, it is now known that some genetically inherited diseases have more severe symptoms each succeeding generation due to segments of the defective genes being doubled in their transmission to children (as illustrated below). These are referred to as stuttering alleles or unstable alleles. Examples of this phenomenon are Huntington's disease, fragile-X syndrome, and the myotonic form of muscular dystrophy.

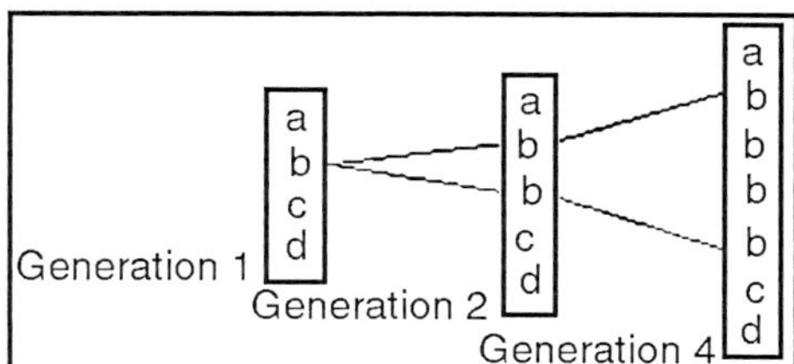

Fig. Unstable Allele Doubling each Generation

Mendel believed that all units of inheritance are passed on to offspring unchanged. Unstable alleles are an important exception to this rule.

Environmental Influences

The phenotype of an individual is not only the result of inheriting a particular set of parental genes. The specific environmental characteristics of the uterus in which a fertilized egg is implanted and the health of the mother can have major impacts on the phenotype of the future child. For instance, oxygen deprivation or inappropriate hormone levels can cause lifelong, devastating effects. Likewise, accidents, poor nutrition, and other environmental influences throughout life can alter an individual's phenotype. Geneticists study identical or monozygotic twins to determine which traits are inherited and which ones were acquired following conception. Since monozygotic twins come from the same zygote, they are essentially identical in their genetic makeup. If there are any differences in their phenotypes, the environment is virtually always responsible. Such differences show up in basic capabilities such as handedness, which had been assumed to be entirely genetically determined. In rare instances, one monozygotic twin will be clearly right-handed while the other will be left-handed. This suggests that there may be both genetic and environmental influences in the development of this trait.

Researchers have identified more than 5,000 genetically inherited human diseases and abnormalities. As we learn more about the inheritance patterns for these traits, it is becoming clear that at least some of the twelve exceptions to the simple Mendelian rules of inheritance described here are, in fact, relatively common. It would not be surprising if other "exceptions" were discovered in the future. However, it is important to keep in mind that there are at least 18,000 human traits controlled by genes that follow the basic Mendelian rules of inheritance.

6

Fats and Lipids

FAT (LIPIDS)

Despite some recent literature wrongly espousing the benefits of high fat intakes (*i.e.*, intakes of 30 Percent or more of total calories from fat), fat is a highly concentrated fuel that does nothing to improve athletic performance, body composition, or weight when taken in excess. The adult AMDR for total fat intake is 20 to 35 Percent of total calories, and there is no scientific information suggesting that more than 25 Percent of total calories from fat is generally better for athletes. However, for athletes who may have difficulty sustaining weight because of a massive energy expenditure (such as cross-country skiers) or who must sustain high weights (such as linemen on football teams), higher fat intakes (up to the AMDR limit of 35 Percent) may be necessary.

Few Americans consume less than 35 Percent of total calories from fat, so consumption of less fat is not easy, and unless steps are taken to provide sufficient energy from other substrates (mainly from more complex carbohydrates) to replace the eliminated fat, athletes may place themselves in an energy-deficit state that is, in itself, a detriment to performance. Therefore, while a reduction in fat intake is generally useful, a conscious effort should be made to provide enough total energy when fat intake is reduced. Since fat is more than twice as concentrated in calories than either protein or carbohydrate (9 calories per gram versus 4 calories per gram), more than twice as much food must be consumed to make up the difference in reduced fat.

Cholesterol, oils, butter, and margarine are all fats, or lipids, but each has slightly different characteristics. The one common attribute shared by lipids is that they are soluble in organic solvents but not soluble in water. (Anyone who has tried to mix Italian dressing knows this to be true. The oil in the dressing eventually rises to the top, no matter how hard the bottle is shaken.) The term fat is usually applied to lipids that are solid at room temperature, and the term oil is applied to lipids that are liquid at room temperature.The most commonly consumed form of lipid is triglyceride, which consists of three fatty acids and one glycerol molecule (thus the name tri glyceride). Despite the numerous forms of lipids, we can obtain them all from the food supply, and we are also capable of making many types of lipids by combining carbon units from other substances. Nearly every cell in the body has the capacity to make cholesterol, which is why a person can have a high blood cholesterol leveleven when on a low-cholesterol diet.

We can also manufacture phospholipids, triglycerides, and oils. In fact, it is this ability to effectively manufacture different types of lipids that limits the necessity to consume large amounts of lipids.

STRUCTURE AND FUNCTION

FAT FUNCTIONS

A certain amount of fat, between 20 and 35 Percent of total consumed calories, is necessary to ensure a sufficient energy and nutrient intake. The fat-soluble vitamins-vitamins A,D,E, and K-must be delivered in a fat package. The essential fatty acids, which are needed for specific body functions but that we are incapable of synthesizing, must also be consumed. Some dietary fat is also needed to give us a feeling of satiety during the meal, creating the important physiological signal that it is time to stop eating.

Dietary fats have a longer gastric emptying time than do carbohydrates, which contributes to the feeling of satiety. Of course, fat also makes foods taste good.

LIPIDS

All Lipids are hydrophobic: that's the one property they have in common. This group of molecules includes fats and oils, waxes, phospholipids, steroids (like cholesterol), and some other related compounds.

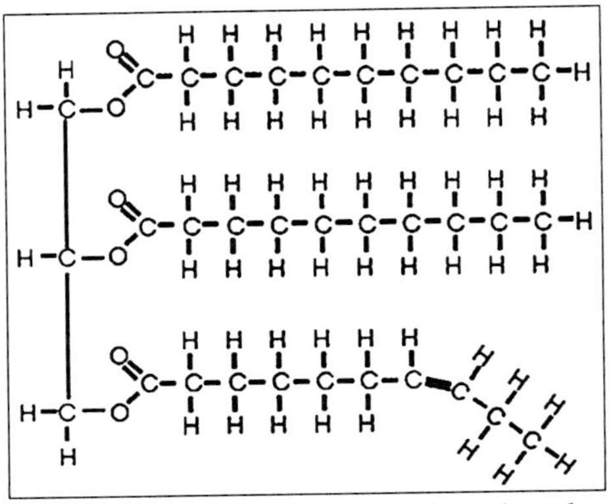

Fats and oils are made from two kinds of molecules: glycerol (a type of alcohol with a hydroxyl group on each of its three carbons) and three fatty acids joined by dehydration synthesis. Since there are three fatty acids attached, these are known as triglycerides."Bread" and pastries from a"bread factory" often contain mono- and diglycerides as"dough conditioners." Can you figure out what these molecules would look like? The main distinction between fats and oils is whether they're solid or liquid at room temperature, and this, as we'll soon see, is based on differences in the structures of the fatty acids they contain.

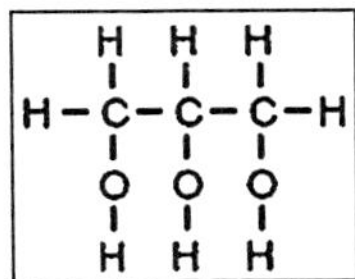

Lipid Structure

Lipids have different levels of saturation, a term that refers to the number of double bonds in the carbon chain. Fatty acids with no double bonds are saturated, those with one double bond are monounsaturated, and those with more than one double bond are polyunsaturated. Single bonds are stronger and less chemically reactive than double bonds, so the greater the number of double bonds, the greater the opportunity for the fatty acid to react with its chemical environment. It is this differential reactive capacity that makes the number of double bonds an important factor in human nutrition.

Saturated fatty acids are most prevalent in fats of animal origin, palm kernel oil, and coconut oil. Monounsaturated fats are highest in olive oil and canola oil but are also present in fats of animal origin. Polyunsaturated fats are highest in vegetable oils (with the exception of olive oil, which is more than 75 Percent monounsaturated). In the context of a fat intake that does not exceed 35 Percent of total calories, mono- and polyunsaturated fats should make up the majority of the fats consumed. Saturated fats are associated with higher cholesterol levels, so they should be minimized when possible. This is most easily achieved by reducing the consumption of animal fats, chocolate candies (often high in saturated tropical oils), fried foods, and high-fat dairy products.

STRUCTURE OF FATTY ACIDS

The"tail" of a fatty acid is a long hydrocarbon chain, making it hydrophobic. The"head" of the molecule is a carboxyl group which is hydrophilic. Fatty acids are the main component of soap, where their tails are soluble in oily dirt and their heads are soluble in water to emulsify and wash away the oily dirt. However, when the head end is attached to glycerol to form a fat, that whole molecule is hydrophobic.

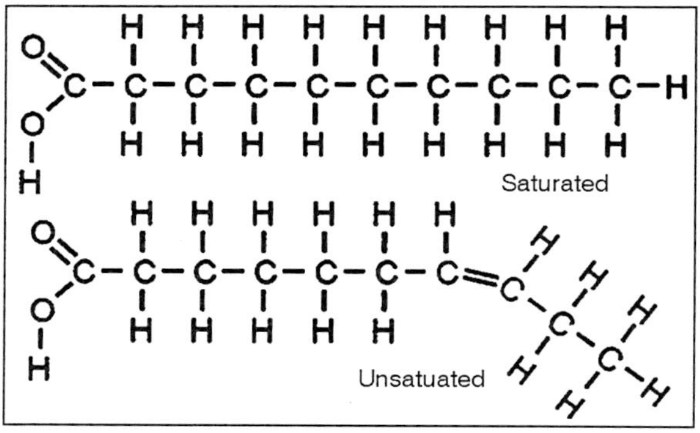

Saturated

Unsatuated

The terms saturated, mono-unsaturated, and poly-unsaturated refer to the number of hydrogens attached to the hydrocarbon tails of the fatty acids as compared to the number of double bonds between carbon atoms in the tail.

Fats, which are mostly from animal sources, have all single bonds between the carbons in their fatty acid tails, thus all the carbons are also bonded to the maximum number of hydrogens possible. Since the fatty acids in these triglycerides contain the maximum possible amouunt of hydrogens, these would be calledsaturated fats.

The hydrocarbon chains in these fatty acids are, thus, fairly straight and can pack closely together, making these fats solid at room temperature. Oils, mostly from plant sources, have some double bonds between some of the carbons in the hydrocarbon tail, causing bends or"kinks" in the shape of the molecules. Because some of the carbons share double bonds, they're not bonded to as many hydrogens as they could if they weren't double bonded to each other. Therefore these oils are called unsaturated fats.

Because of the kinks in the hydrocarbon tails, unsaturated fats can't pack as closely together, making them liquid at room temperature.

Many people have heard that the unsaturated fats are"healthier" than the saturated ones. Hydrogenated vegetable oil (as in shortening and commercial peanut butters where a solid consistency is sought) started out as"good" unsaturated oil.

However, this commercial product has had all the double bonds artificially broken and hydrogens artificially added (in a chemistry lab-type setting) to turn it into saturated fat that bears no resemblance to the original oil from which it came (so it will be solid at room temperature).

In unsaturated fatty acids, there are two ways the pieces of the hydrocarbon tail can be arranged around a C=C double bond.

In cis bonds, the two pieces of the carbon chain on either side of the double bond are either both"up" or both"down," such that both are on the same side of the molecule. In trans bonds, the two pieces of the molecule are on opposite sides of the double bond, that is, one"up" and one"down" across from each other.

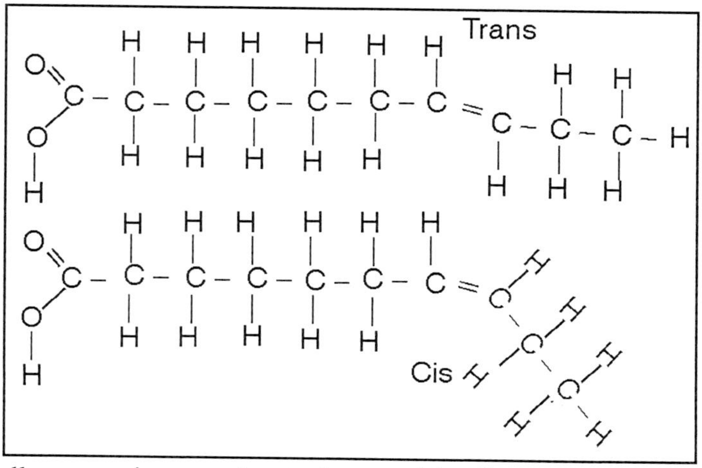

Trans

Cis

Naturally-occurring unsaturated vegetable oils have almost all cis bonds, but using oil for frying causes some of the cis bonds to convert to trans bonds. If oil is used only once like when you fry an egg, only a few of the bonds do this so it's not too bad.

However, if oil is constantly reused, like in fast food French fry machines, more and more of the cis bonds are changed to trans until significant numbers of fatty acids with trans bonds build up. The reason this is of concern is that fatty acids with trans bonds are carcinogenic, or cancer-causing.

The levels of trans fatty acids in highly-processed, lipid-containing products such as margarine are quite high, and I have heard that the government is considering requiring that the amounts of trans fatty acids in such products be listed on the labels.

We need fats in our bodies and in our diet. Animals in general use fat for energy storage because fat stores 9 KCal/g of energy. Plants, which don't move around, can afford to store food for energy in a less compact but more easily accessible form, so they use starch (a carbohydrate, NOT A LIPID) for energy storage. Carbohydrates and proteins store only 4 KCal/g of energy, so fat stores over twice as much energy/gram as fat. By the way, this is also related to the idea behind some of the high-carbohydrate weight loss diets.

The human body burns carbohydrates and fats for fuel in a given proportion to each other. The theory behind these diets is that if they supply carbohydrates but not fats, then it is hoped that the fat needed to balance with the sugar will be taken from the dieter's body stores.

Fat is also is used in our bodies to a) cushion vital organs like the kidneys and b) serve as insulation, especially just beneath the skin.

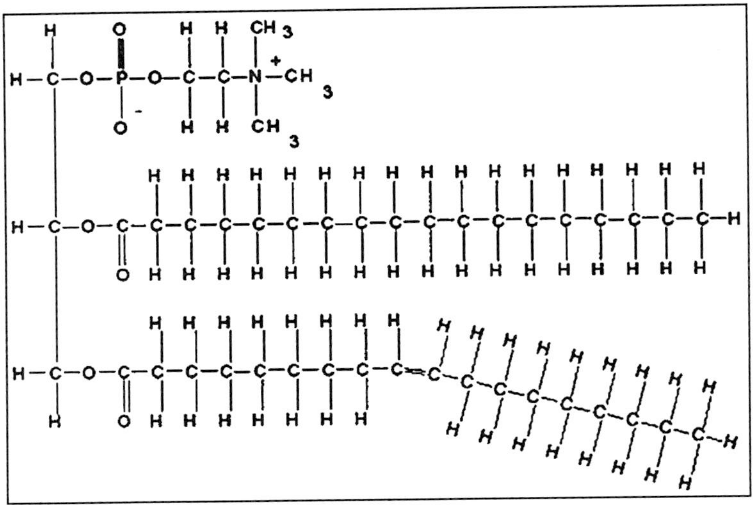

PHOSPHOLIPIDS

Phospholipids are made from glycerol, two fatty acids, and (in place of the third fatty acid) a phosphate group with some other molecule attached to its other end. The hydrocarbon tails of the fatty acids are still hydrophobic, but the phosphate group end of the molecule is hydrophilic because of the oxygens with all of their pairs of unshared electrons. This means that phospholipids are soluble in both water and oil.

An emulsifying agent is a substance which is soluble in both oil and water, thus enabling the two to mix. A"famous" phospholipid is lecithin which is found in egg yolk and soybeans.

Egg yolk is mostly water but has a lot of lipids, especially cholesterol, which are needed by the developing chick. Lecithin is used to emulsify the lipids and hold them in the water as an emulsion. Lecithin is the basis of the classic emulsion known as mayonnaise.

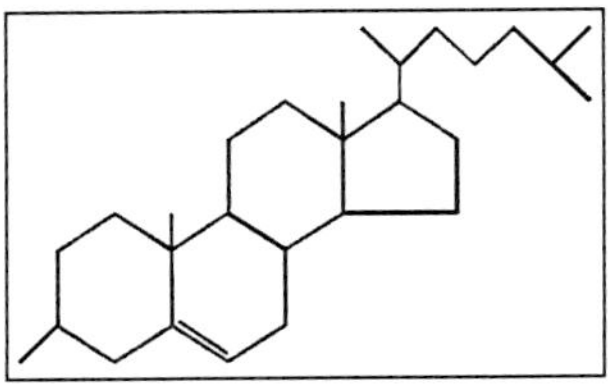

Our cell membranes are made mostly of phospholipids arranged in a double layer with the tails from both layers"inside" (facing Towards each other) and the heads facing"out" (Towards the watery environment) on both surfaces.

STEROIDS

The general structure of cholesterol consists of two six-membered rings side-by-side and sharing one side in common, a third six-membered ring off the top corner of the right ring, and a five-membered ring attached to the right side of that. The central core of this molecule, consisting of four fused rings, is shared by all steroids, including estrogen (estradiol), progesterone, corticosteroids such as cortisol (cortisone), aldosterone, testosterone, and Vitamin D. In the various types of steroids, various other groups/molecules are attached around the edges. Know how to draw the four rings that make up the central structure.

Cholesterol is not a"bad guy!" Our bodies make about 2 g of cholesterol per day, and that makes up about 85per cent of blood cholesterol, while only about 15per cent comes from dietary sources. Cholesterol is the precursor to our sex hormones and Vitamin D. Vitamin D is formed by the action of UV light in sunlight on cholesterol molecules that have"risen" to near the surface of the skin. At least one source I read suggested that people not shower immediately after being in the sun, but wait at least ½ hour for the new Vitamin D to be absorbed deeper into the skin. Our cell membranes contain a lot of cholesterol (in between the phospholipids) to help keep them"fluid" even when our cells are exposed to cooler temperatures.

Many people have hear the claims that egg yolk contains too much cholesterol, thus should not be eaten. An interesting study was done at Purdue University a number of years ago to test this. Men in one group each ate an egg a day, while men in another group were not allowed to eat eggs. Each of these groups was further subdivided such that half the men got"lots" of exercise while the other half were"couch potatoes." The results of this experiment showed no significant difference in blood cholesterol levels between egg-eaters and non-egg-eaters while there was a very significant difference between the men who got exercise and those who didn't.

Lipoproteins are clusters of proteins and lipids all tangled up together. These act as a means of carrying lipids, including cholesterol, around in our blood. There are two main categories of lipoproteins distinguished by how compact/dense they are. LDL or low density lipoprotein is the"bad guy," being associated with deposition of"cholesterol" on the walls of someone's arteries. HDL or high density lipoprotein is the"good guy," being associated with carrying"cholesterol" out of the blood system, and is more dense/more compact than LDL.

Triglycerides

The majority of consumed lipids are triglycerides, which contain three fatty acids and a glycerol molecule. Fat is stored in the form of triglycerides, which we manufacture when excess energy is consumed. We store triglycerides in adipose tissues (groups of fat cells) and inside muscle cells (intramuscular tri glyceride), both of which are available as an energy source when needed. When fat is burned as a source of energy, the stored triglycerides are taken out of storage, and each molecule is cleaved into its component fatty acids and glycerol molecule. Each fatty acid can then be broken apart (two carbon units at a time) and thrown into the cellular furnaces for the creation of ATP to form heat and provide the energy for muscular work. This process is referred to as the beta-oxidative metabolic pathway because burning fat, besides requiring some carbohydrate for its complete oxidation, also requires oxygen.

Glycerol is a unique lipid that is burned like a carbohydrate rather than a fat and is also an effective humectant (it holds water). Some long-endurance athletes find that adding glycerol to water helps them retain more water (*i.e.*, to superhydrate) than if they consumed water alone. In an extremely hot and humid environment, water loss is likely to be higher than the athlete's fluid-replacement capacity, so beginning a competition in a superhydrated state may provide some advantages.

In studies of tennis and Olympic distance triathlon, athletes consuming glycerol in water prior to competition experienced some protective hyperhydration benefits when exercising in high heat.17,18 However, being superhydrated is likely to impart a degree of discomfort that requires some adaptation. Athletes who consume fluids containing glycerol often describe the feeling of holding extra body water as making them feel"like a water bag," or"heavy," or"stiff." Still, they maintain that how they feel at the end of a race is more important than how they feel at the beginning, so adding glycerol to precompetition fluids has become a standard protocol for some athletes.

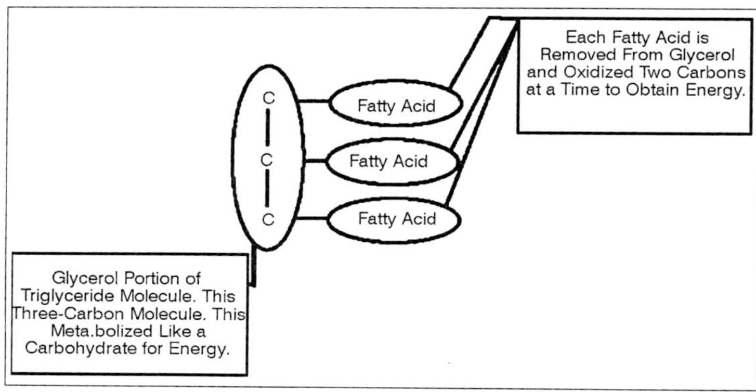

FATTY ACIDS

Linoleic (omega-6) and linolenic (omega-3) fatty acids are the essential fatty acids; although they are needed for metabolic processes, we are incapable of synthesizing them. The"omega-6" classification means that these fatty acids, which are polyunsaturated, have the last double bond six carbons from the end of the carbon chain.

The"omega-3" classification means that these fatty acids have the last double bond three carbons from the end of the carbon chain. Linoleic acid is an essential part of lipid membranes and is required for normal skin health. Linolenic acid is necessary for neural function and growth. The AMDR for omega-6 andomega-3 fatty acids is 5 to 10 and.6 to 1.2 grams per day, respectively.11Both fatty acids are easily obtained from vegetable oils (corn, safflower, canola, and so on) and the oils of cold-water fish.

Fish liver oils that are high in omega-3 acids have recently received much attention. These oils have been shown to reduce the ability of red blood cells to congregate, thereby decreasing the chance for an unwanted blood clot to form. This reduces the risk of a heart attack, which is most commonly caused by a clot formation in one of the major heart arteries. The oils from cold-water fish are the main source of the omega-3 fatty acids eicosapentanoic acid (EPA) and docosohexanoic acid (DHA).

Even a once-weekly consumption of cold-water fish (salmon, albacore tuna, Atlantic herring) is sufficient to significantly reduce the risk of heart attack and stroke.11Despite these findings, excessive intake of these fish oils may cause problems, including an increase in cellular oxidative damage. The best rule of thumb is to make fish consumption a regular part of the weekly diet so that supplemental intake of omega-3 fatty acids is unnecessary.

FAT REQUIREMENTS

From an exercise standpoint, there is little reason to believe that increasing fat consumption results in improving athletic performance, unless the increase in fat intake is the only reasonable means for the athlete to obtain sufficient energy. For the athlete who needs more than 4,000 calories each day to meet the combined demands of growth, exercise, and maintenance, moderate increases in dietary fat (preferably from plant and fish sources) may be needed. Since fat is a more concentrated form of energy than either carbohydrate or protein, more energy can be consumed in a smaller food package if the foods contain more fat.

LIPIDS AND PHYSICAL ACTIVITY

Even the leanest, healthiest athletes have a substantial energy pool of stored lipids. The average storage in adipose tissue ranges between 50,000 and 100,000 calories, or enough energy to walk or run 500 to 1,000 miles19 (800 to 1,600 kilometers) without a refueling stop. In addition, athletes store approximately 2,000 to 3,000 calories of lipids inside the muscle tissue.20 These lipids, which are stored in the form of triglycerides, are available as a fuel under the proper conditions of oxygen availability. Maximal fat oxidation occurs at 60 to 65 Percent VO_2 max, but at higher levels of VO2max there is insufficient oxygen to derive the majority of energy used from fat catabolism.

Triglycerides stored in adipose tissue are cleaved into their component molecules of glycerol and fatty acids and transported to the blood plasma. The glycerol is available to all tissues for energy metabolism, and the free fatty acids are transported to working muscles where they are oxidized for energy. The triglycerides stored in working muscles are cleaved into glycerol and fatty acids, and the fatty acids can be oxidized for energy where they are resident. The glycerol can also be burned for energy in the working muscle or can be transported to the blood plasma as a source of energy for other tissues.

The lower the exercise intensity, the greater the proportion of fat burned to satisfy energy needs. As exercise intensity increases, the proportion of fat burned decreases and the proportion of carbohydrate burned increases. It is this basic reality that is behind why so many people do low-intensity activity to burn fat and lower body fat levels. However, the proportion of fat burned should not be confused by the total amount of fat burned at different intensities of physical activity. As exercise intensity increases, the total number of calories burned per unit of time also increases. Although there may be a decrease in the proportion of fat burned to satisfy total energy needs in higher-intensity activity, the total volume of fat burned is greater because the total energy requirement is higher.

The take-away message from this metabolic reality is that athletes interested in lowering body fat should exercise at least as high as 65 Percent of VO2 max for the duration of their workouts to optimize the total mass of fat that is burned. Exercising at lower intensities burns a greater proportion of fat but less total fat than exercising at higher intensities.

BIOSYNTHESIS OF FATTY ACID

Identification of Genes in Biosynthesis of Fatty Acids and Cholesterol Pathways Using Microarrays Sterol regulatory binding proteins (SREBP) are a family of basic helix-loop-helix leucine zipper transcription factors that regulate biosynthesis of fatty acids and cholesterol. To identify candidate genes in the fatty acid synthesis pathway, transgenic mice over expressing SREBP-1a and SREBP-2 isofroms in liver were used. Labeled cRNA was prepared from liver and hybridized to mouse expression microarrays interrogating ~30,000 genes/ ESTs. Some of the known genes, ACC, FAS, and ATP citrate lyase regulated by over expression of SREBP were identified by microarray analysis. A novel gene, long chain fatty acyl elongase (LCE) regulated by the SREBP1a and 2 was also identified. LCE was over expressed with a fold change of 25 and 9 in SREBP-1a and SREBP-2 mice respectively. The confirmation of LCE was done by northern blot analysis, sequence homology searches and biochemical tests. This enzyme specifically catalyzes the elongation of 12, 14 and 16 carbons but does not elongate the C-18 fatty acids or very long chain fatty acids.

DEGRADATION OF FATTY ACIDS

Multi-million pounds of viscera are generated from the catfish filleting process. Our previous study showed that catfish viscera contained 30-35per cent crude fat, and catfish visceral oil contained health-promoting fatty acids. Catfish visceral oils show potential for use as nutrient supplement and as cooking oil. Information on thermal stability and degradation of catfish visceral oils is not available.

The objective of this study was to determine weight loss (per cent) of catfish visceral oils and fatty acids after having been subjected to heat treatment. Oil samples included crude, degummed, neutralized, bleached, and deodorized oils.

The oils and individual fatty acids, FAs (C14:0, C16:0, C18:0, C20:0, C16:1, C18:1, C18:2, C18:3, C20:2, C20:4, C22:6) were subjected to the Thermogravimetric Analyser. Samples (0.5-1 mg) was placed in the furnace and heated up to 600°C at the rate of 5°C/min. Weight loss (per cent) of oils and fatty acids were calculated based on the original weight of non-heated samples. Duplicate experiments were conducted.

Over 90per cent weight loss of all FAs (except C18:1 and C20:2) was observed at 250°C. Less than 2per cent remaining weight of FAs (except C20:2) was observed at 350°C. Saturated FAs completely decomposed above 400°C. Comparing C18:1 with C18:2 and C18:3, weight loss increased with increased double bonds. Between 200-550°C, weight loss of oils slightly increased with increased temperature, regardless of the purification process. There was, however, no distinct weight loss of all oils below 300°C. Oils with less impurity had greater weight loss especially at 450-500°C. At 550°C, all oil samples completely decomposed.

Knowledge of thermal stability or degradation (weight loss through eVapouration or decomposition) of catfish oils during purification or cooking is essential for the design of processing units, analysis of production cost, and cooking quality if to be used as cooking or seasoning oil.

CHOLESTROL BIOSYNTHESIS AND DEGRADATION

CHOLESTEROL BIOSYNTHESIS AND CATABOLISM

The formation of cholesterol by the liver, regulated by a type of homeostatic control, is influenced by the hepatic level offree cholesterol or glycogen, or both, and by the availability of the proper enzymes. Here the factors affecting the hepaticbiosynthesis of cholesterol are detailed, including the influence of dietary composition and fasting.

The specific effects oncholesterol synthesis of feeding cholesterol, various steroids, saturated and unsaturated fats and thyroid hormone are reviewed. Also considered are the effects of radiation, glycolysis and various pharmacologic agents. In the degradation of cholesterol, the chief metabolic end products are bile acids, although cholesterol may also be converted into sex hormones and adrenocortical hormones. The steps involved in the degradation of cholesterol to bileacids are reviewed, along with the various factors which increase or decrease cholesterol catabolism. Since degradation of cholesterol depends on the levels of bile salts in the liver, any factor which will increase the turnover of bile acids will also accelerate cholesterol catabolism and may in turn lower serum cholesterol.

CHOLESTEROL DESCRIPTION

"Cholesterol is a vital constituent of cell membranes and the precursor for steroid hormones and bile acids. It is clearly essential to life, yet its deposition in arteries is associated with cardiovascular disease and stroke, two leading causes of death in humans. In a healthy organism, an intricate balance is maintained between the biosynthesis, utilization, and transport of cholesterol, keeping its harmful deposition to a minimum."

HOLESTEROL A CRUCIAL COMPONENT OF CELL MEMBRANES

- Cholesterol is the third major lipid in membranes.
- Cholesterol is compact,rigid,hydrophobicmolecule with a polar hydroxyl group.
- The hydroxyl group of cholesterol is orientedtowards the aqueous phase (polar heads of phospholipids) and gives hydrophilic character to this end of cholesterol molecule.
- The cholesterol to polar lipid ratios affect stability, fluidity, permeability, and protein mobility. Cholesterol also controls the microstructure of membranes.
- Cholesterol is especially abundant in myelinated structures of the brain and central nervous system and almost absent from the inner mitochondrial membrane.

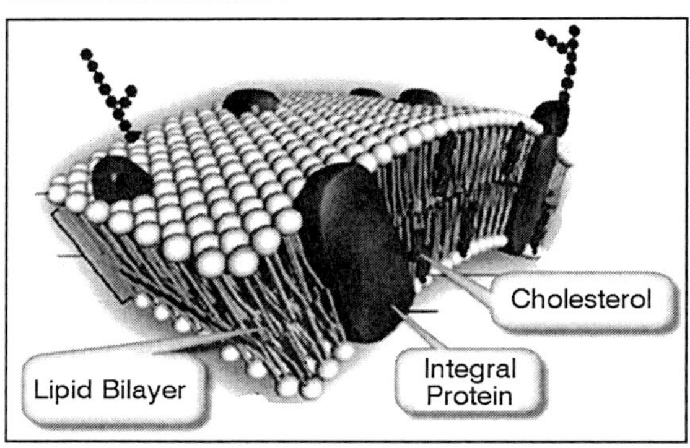

CHOLESTEROL IS PRECURSOR OF BILE SALTS

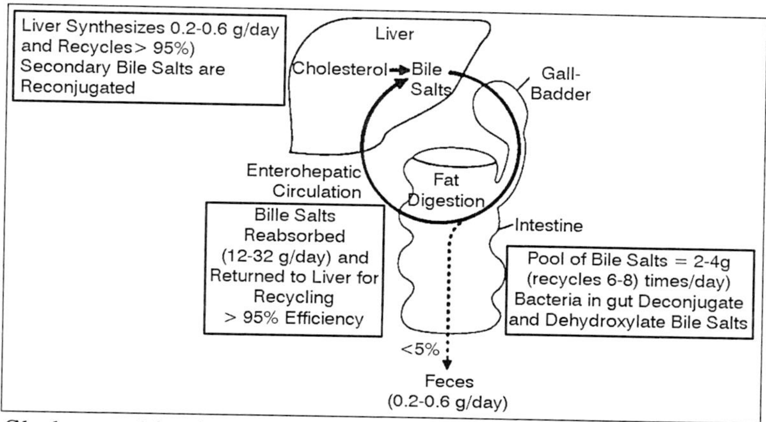

- Cholesterol is abundant in bile.
- The solubilization of hydrophobic molecules of cholesterol is aided by bile phospholipids and bile salts. They prevent cholesterol from precipitating in gallbladder in form of gallstones.
- Bile salts are metabolites of cholesterol.
- Cholesterol protects gallbladder membranes from irritating and harmful effects of bile acids and bile salts.

Cholesterol is the precursor of steroid hormones

Testosterone: A Male Sex Hormone
Estradol: a Female Sex Hormone
Estrone: A Female Sex Hormone

Cholesterol is the precursor of ALL steroid hormones including progesterone, estrogen, testosterone,corticosteroids, and aldosterone.

Thus, cholesterol is crucial for human reproduction, response to the environment, and preservation of human species.

CHOLESTEROL IS PRECURSOR OF VITAMIN D

Active Vitamin D_3
Cholecalciferol
Cholesterol
7-Ddehydrocholesterol
UV Light
Previtamin D_3

- The various forms of vitamin D, which are really hormones, aresterol derivatives.
- Active vitamin D increases Ca2+concentration in serum by promoting the intestinal absorption of dietary Ca2+. This increases the deposition of Ca2+ in bones and teeth.

BIOSYNTHESIS OF CHOLESTEROL

Slightly less than half of the cholesterol in the body derives from biosynthesis de novo. Biosynthesis in the liver accounts for approximately 10per cent, and in the intestines approximately 15per cent, of the amount produced each day. Cholesterol synthesis occurs in the cytoplasm and microsomes from the two-carbon acetate group of acetyl-CoA.

The process has five major steps:

- Acetyl-CoAs are converted to 3-hydroxy-3-methylglutaryl-CoA (HMG-CoA).
- HMG-CoA is converted to mevalonate.
- Mevalonate is converted to the isoprene based molecule, isopentenyl pyrophosphate (IPP), with the concomitant loss of CO_2.
- IPP is converted to squalene.
- Squalene is converted to cholesterol.

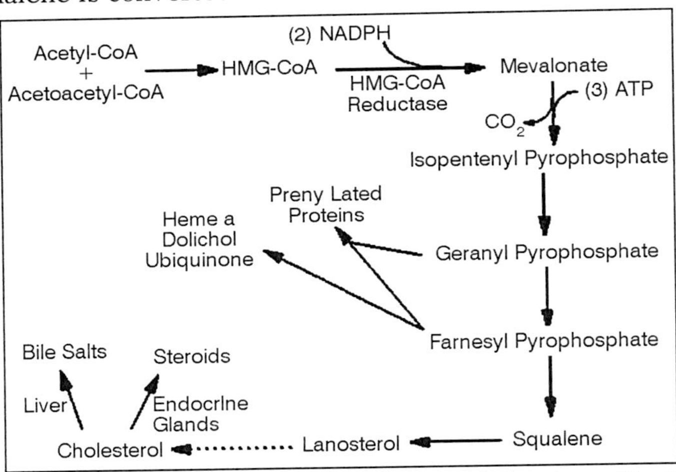

Pathway of cholesterol biosynthesis. Synthesis begins with the transport of acetyl-CoA ffrom the mitochondrion to the cytosol. The rate limiting step occurs at the 3-hydroxy-3

methlgltarlCoA(HMGCoA)redcataseHMGRcataledstepThe phosphorylation reactions are required to solubilize the isoprenoid intermediates in the pathway. Intermediates in the pathway are used for the synthesis of prenylated proteins, dolichol, coenzyme Q and the side chain of heme a. The acetyl-CoA utilized for cholesterol biosynthesis is derived from an oxidation reaction (eg, fatty acids or pyruvate) in the mitochondria and is transported to the cytoplasm by the same process as that described for fatty acid synthesis. Acetyl-CoA can also be derived from cytoplasmic oxidation of ethanol by acetyl-CoA synthetase. All the reduction reactions of cholesterol biosynthesis use NADPH as a cofactor. The isoprenoid intermediates of cholesterol biosynthesis can be diverted to other synthesis reactions, such as those for dolichol (used in the synthesis of N-linked glycoproteins, coenzyme Q (of the oxidative phosphorylation) pathway or the side chain of heme a. Additionally, these intermediates are used in the lipid modification of some proteins.

Acetyl-CoA units are converted to mevalonate by a series of reactions that begins with the formation of HMG-CoA. Unlike the HMG-CoA formed during ketone body synthesis in the mitochondria, this form is synthesized in the cytoplasm. However, the pathway and the necessary enzymes are the same as those in the mitochondria. Two moles of acetyl-CoA are condensed in a reversal of the thiolase reaction, forming acetoacetyl-CoA.

Acetoacetyl-CoA and a third mole of acetyl-CoA are converted to HMG-CoA by the action of HMG-CoA synthase. HMG-CoA is converted to mevalonate by HMG-CoA reductase, HMGR (this enzyme is bound in the endoplasmic reticulum, ER). HMGR absolutely requires NADPH as a cofactor and two moles of NADPH are consumed during the conversion of HMG-CoA to mevalonate.The reaction catalyzed by HMGR is the rate limiting step of cholesterol biosynthesis, and this enzyme is subject to complex regulatory controls.

Mevalonate is then activated by three successive phosphorylations, yielding 5-pyrophosphomevalonate. In addition to activating mevalonate, the phosphorylations maintain its solubility, since otherwise it is insoluble in water. After phosphorylation, an ATP-dependent decarboxylation yields isopentenyl pyrophosphate, IPP, an activated isoprenoid molecule. Isopentenyl pyrophosphate is in equilibrium with its isomer, dimethylallyl pyrophosphate, DMPP.

One molecule of IPP condenses with one molecule of DMPP to generate geranyl pyrophosphate, GPP. GPP further condenses with another IPP molecule to yield farnesyl pyrophosphate, FPP. Finally, the NADPH-requiring enzyme, squalene synthase catalyzes the head-to-tail condensation of two molecules of FPP, yielding squalene (squalene synthase also is tightly associated with the endoplasmic reticulum). Squalene undergoes a two step cyclization to yield lanosterol.

The first reaction is catalyzed by squalene monooxygenase. This enzyme uses NADPH as a cofactor to introduce molecular oxygen as an epoxide at the 2,3 position of squalene. Through a series of 19 additional reactions, lanosterol is converted to cholesterol.

REGULATING CHOLESTEROL SYNTHESIS

Normal healthy adults synthesize cholesterol at a rate of approximately 1g/day and consume approximately 0.3g/day. A relatively constant level of cholesterol in the body (150 - 200 mg/dL) is maintained primarily by controlling the level of de novo synthesis. The level of cholesterol synthesis is regulated in part by the dietary intake of cholesterol. Cholesterol from both diet and synthesis is utilized in the formation of membranes and in the synthesis of the steroid hormones and bile acids. The greatest proportion of cholesterol is used in bile acid synthesis.

The cellular supply of cholesterol is maintained at a steady level by three distinct mechanisms:

- Regulation of HMGR activity and levels
- Regulation of excess intracellular free cholesterol through the activity of acyl-CoA:cholesterol acyltransferase, ACAT
- Regulation of plasma cholesterol levels via LDL receptor-mediated uptake and HDL-mediated reverse transport.

Regulation of HMGR activity is the primary means for controlling the level of cholesterol biosynthesis.

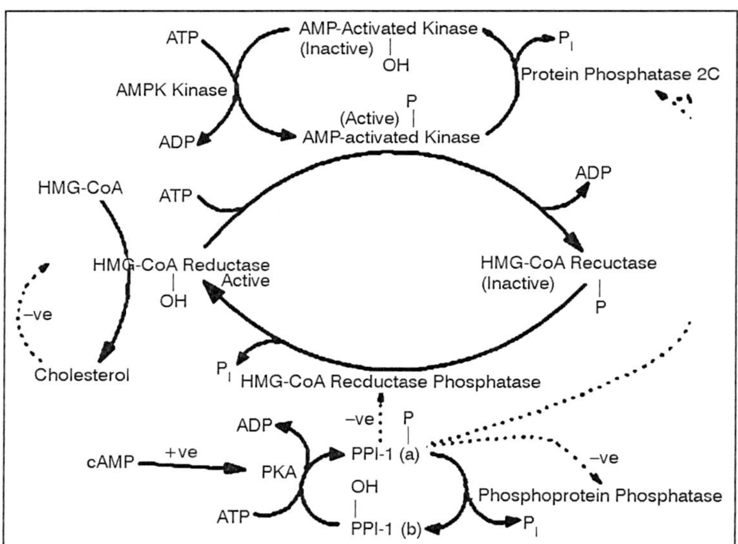

The enzyme is controlled by four distinct mechanisms: feed-back inhibition, control of gene expression, rate of enzyme degradation and phosphorylation-dephosphorylation.The first three control mechanisms are exerted by cholesterol itself. Cholesterol acts as a feed-back inhibitor of pre-existing HMGR as well as inducing rapid degradation of the enzyme. The latter is the result of cholesterol-induced polyubiquitination of HMGR and its degradation in the proteosome (see proteolytic degradation below). This ability of cholesterol is a consequence of the sterol sensing domain, SSD of HMGR. In addition, when cholesterol is in excess the amount of mRNA for HMGR is reduced as a result of decreased expression of the gene. The mechanism by which cholesterol (and other sterols) affect the transcription of the HMGR gene is described below under regulation of sterol content.

Regulation of HMGR through covalent modification occurs as a result of phosphorylation and dephosphorylation. The enzyme is most active in its unmodified form. Phosphorylation of the enzyme decreases its activity. HMGR is phosphorylated by AMP-activated protein kinase, AMPK (this is not the same as cAMP-dependent protein kinase, PKA). AMPRK itself is activated via phosphorylation. The phosphorylation of AMPK is catalyzed by AMPK kinase.

Regulation of HMGR by covalent modification. HMGR is most active in the dephosphorylated state. Phosphorylation is catalyzed by AMP-activated protein kinase, AMPK, (used to be termed HMGR kinase), an enzyme whose activity is also regulated by phosphorylation. Phosphorylation of AMPK is catalyzed by AMPK kinase. Hormones such as glucagon and epinephrine negatively affect cholesterol biosynthesis by increasing the activity of the inhibitor of phosphoprotein phosphatase inhibitor-1, PPI-1. Conversely, insulin stimulates the removal of phosphates and, thereby, activates HMGR activity. Additional regulation of HMGR occurs through an inhibition of its' activity as well as of its' synthesis by elevation in intracellular cholesterol levels. This latter phenomenon involves the transcription factor SREBP described. The activity of HMGR is additionally controlled by the cAMP signaling pathway. Increases in cAMP lead to activation of cAMP-dependent protein kinase, PKA. In the context of HMGR regulation, PKA phosphorylates phosphoprotein phosphatase inhibitor-1 (PPI-1) leading to an increase in its' activity. PPI-1 can inhibit the activity of numerous phosphatases including protein phosphatase 2C and HMG-CoA reductase phosphatase which remove phosphates from AMPRK and HMGR, respectively. This maintains AMPRK in the phosphorylated and active state, and HMGR in the phosphorylated and inactive state. As the stimulus leading to increased cAMP production is removed, the level of phosphorylations decreases and that of dephosphorylations increases. The net result is a return to a higher level of HMGR activity.

Since the intracellular level of cAMP is regulated by hormonal stimuli, regulation of cholesterol biosynthesis is hormonally controlled. Insulin leads to a decrease in cAMP, which in turn activates cholesterol synthesis. Alternatively, glucagon and epinephrine, which increase the level of cAMP, inhibit cholesterol synthesis.

The ability of insulin to stimulate, and glucagon to inhibit, HMGR activity is consistent with the effects of these hormones on other metabolic pathways. The basic function of these two hormones is to control the availability and delivery of energy to all cells of the body. Long-term control of HMGR activity is exerted primarily through control over the synthesis and degradation of the enzyme. When levels of cholesterol are high, the level of expression of the HMGR gene is reduced. Conversely, reduced levels of cholesterol activate expression of the gene. Insulin also brings about long-term regulation of cholesterol metabolism by increasing the level of HMGR synthesis.

Proteolytic Regulation of HMG-CoA Reductase

The stability of HMGR is regulated as the rate of flux through the mevalonate synthesis pathway changes. When the flux is high the rate of HMGR degradation is also high. When the flux is low, degradation of HMGR decreases. This phenomenon can easily be observed in the presence of the statin drugs.

HMGR is localized to the ER and like SREBP contains a sterol-sensing domain, SSD. When sterol levels increase in cells there is an concommitant increase in the rate of HMGR degradation. The degradation of HMGR occurs within the proteosome, a multiprotein complex dedicated to protein degradation. The primary signal directing proteins to the proteosome is ubiquitination. Ubiquitin is a 7.6kDa protein that is covalently attached to proteins targeted for degradation by ubiqitin ligases. These enzymes attach multiple copies of ubiquitin allowing for recognition by the proteosome. HMGR has been shown to be ubiquitinated prior to its degradation. The primary sterol regulating HMGR degradation is cholesterol itself. As the levels of free cholesterol increase in cells, the rate of HMGR degradation increases.

UTILIZATION OF CHOLESTEROL

Cholesterol is transported in the plasma predominantly as cholesteryl esters associated with lipoproteins. Dietary cholesterol is transported from the small intestine to the liver within chylomicrons. Cholesterol synthesized by the liver, as well as any dietary cholesterol in the liver that exceeds hepatic needs, is transported in the serum within LDLs. The liver synthesizes VLDLs and these are converted to LDLs through the action of endothelial cell-associated lipoprotein lipase.

Cholesterol found in plasma membranes can be extracted by HDLs and esterified by the HDL-associated enzyme LCAT. The cholesterol acquired from peripheral tissues by HDLs can then be transferred to VLDLs and LDLs via the action of cholesteryl ester transfer protein (apo-D) which is associated with HDLs. Reverse cholesterol transport allows peripheral cholesterol to be returned to the liver in LDLs. Ultimately, cholesterol is excreted in the bile as free cholesterol or as bile salts following conversion to bile acids in the liver.

REGULATION OF CELLULAR STEROL CONTENT

The continual alteration of the intracellular sterol content occurs through the regulation of key sterol synthetic enzymes as well as by altering the levels of cell-surface LDL receptors. As cells need more sterol they will induce their synthesis and uptake, conversely when the need declines synthesis and uptake are decreased. Regulation of these events is brought about primarily by sterol-regulated transcription of key rate limiting enzymes and by the regulated degradation of HMGR. Activation of transcriptional control occurs through the regulated cleavage of the membrane-bound transcription factor sterol regulated element binding protein,

Sterol control of transcription affects more than 30 genes involved in the biosynthesis of cholesterol, triacylglycerols, phospholipids and fatty acids. Transcriptional control requires the presence of an octamer sequence in the gene termed the sterol regulatory element, SRE-1. It has been shown that SREBP is the transcription factor that binds to SRE-1 elements. It turns out that there are 2 distinct SREBP gene, SREBP-1 and SREBP-2. In addition, the SREBP-1 gene encodes 2 proteins, SREBP-1a and SREBP-1c as a consequence of alternative exon usage. All 3 proteins are proteolytically regulated by sterols. Full-length SREBPs have several domains and are embedded in the membrane of the endoplasmic reticulum (ER). The N-terminal domain contains a transcription factor motif of the basic helix-loop-helix (bHLH) type that is exposed to the cytoplasmic side of the ER. There are 2 transmembrane spanning domains followed by a large C-terminal domain also exposed to the cytosolic side. When sterols are scarce, cleavage of the full-length SREBP takes place with the result being that the N-terminal bHLH motif is released into the cytosol. The bHLH domain then migrates to the nucleus to direct transcription. Conversely, when sterols are abundant, cleavage of SREBP is inhibited. To control the level of SREBP-mediated transcription, the soluble bHLH domain is itself subject to rapid proteolysis.

The cleavage of SREBP is carried out by 2 distinct enzymes, one of which is regulated by sterols. The regulated cleavage occurs in the lumenal loop between the 2 transmembrane domains. This cleavage is catalyzed by site-1 protease, S1P. High sterol content blocks the activity of S1P. The second cleavage, catalyzed by site-2 protease, S2P, occurs in the first transmembrane span, leading to release of active SREBP.

Additional studies on sterol-regulated gene expression demonstrated that cleavage of SREBP by S1P is controlled by the level and action of an additional protein termed, SREBP cleavage-activating protein, SCAP. SCAP is a large protein also found in the ER membrane and contains at least 8 transmembrane spans. The C-terminal portion, which extends into the cytosol, has been shown to interact with the C-terminal domain of SREBP. This C-terminal region of SCAP contains 4 motifs called WD40 repeats.

The WD40 repeats are required for interaction of SCAP with SREBP. Interestingly, the N-terminus of SCAP, including membrane spans 2-6, resembles HMGR which itself is subject to sterol-stimulated degradation (see above). This shared motif is called the sterol sensing domain, SSD. Several proteins whose functions involve sterols also contain the SSD. These include patched, an important development regulating receptor whose ligand, hedgehog, is modified by attachment of cholesterol and the Neimann Pick C1 (NPC1) protein which is involved in cholesterol transport in the secretory pathway. NPC1 is one of several genes whose activities, when disrupted, lead to severe neurological dysfunction.

The function of SCAP is to positively stimulate S1P-mediated cleavage of SREBP. The function of sterols is to inhibit this positive action of SCAP. The activity of SCAP involves movement from the ER to the Golgi and back. Because the C-terminus of SCAP interacts with SREBP, movement of SCAP takes SREBP along for the ride. When sterols are low, SCAP and SREBP move to the Golgi. This transit is required for SREBP cleavage as S1P is Golgi-localized. When sterols are high, movement of SCAP is halted. Thus, the overall effect of sterols is to regulate the ability of SCAP to present SREBP to S1P.

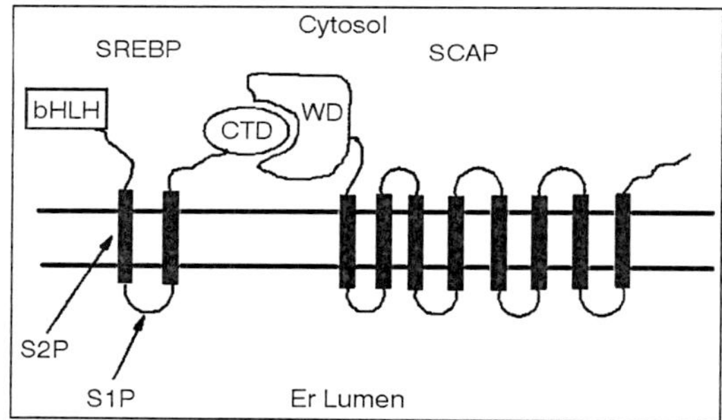

Diagramatic representation of the interactions between SREBP and SCAP in the membrane of the ER CTD = C-terminal domain

BILE ACIDS SYNTHESIS AND UTILIZATION

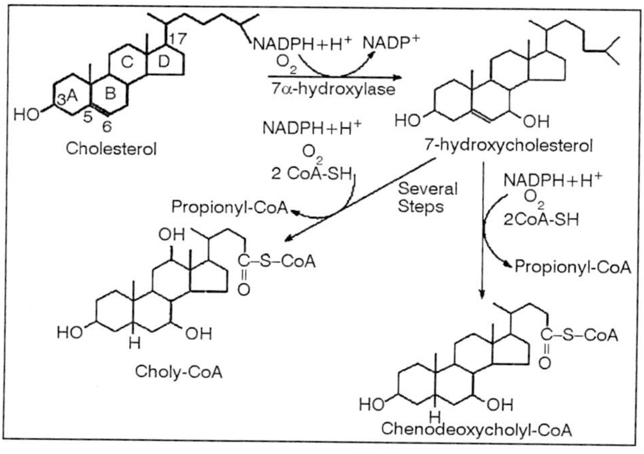

The end products of cholesterol utilization are the bile acids, synthesized in the liver. Synthesis of bile acids is one of the predominant mechanisms for the excretion of excess cholesterol. However, the excretion of cholesterol in the form of bile acids is insufficient to compensate for an excess dietary intake of cholesterol. Synthesis of the 2 primary bile acids, cholic acid and chenodeoxycholic acid. The reaction catalyzed by the 7α-hydroxylase is the rate limiting step in bile acid synthesis. Conversion of 7α-hydroxycholesterol to the bile acids requires several steps not shown in detail in this image. Only the relevant co-factors needed for the synthesis steps are shown.

The most abundant bile acids in human bile are chenodeoxycholic acid (45per cent) and cholic acid (31per cent). These are referred to as the primary bile acids. Within the intestines the primary bile acids are acted upon by bacteria and converted to the secondary bile acids, identified as deoxycholate (from cholate) and lithocholate (from chenodeoxycholate). Both primary and secondary bile acids are reabsorbed by the intestines and delivered back to the liver via the portal circulation.

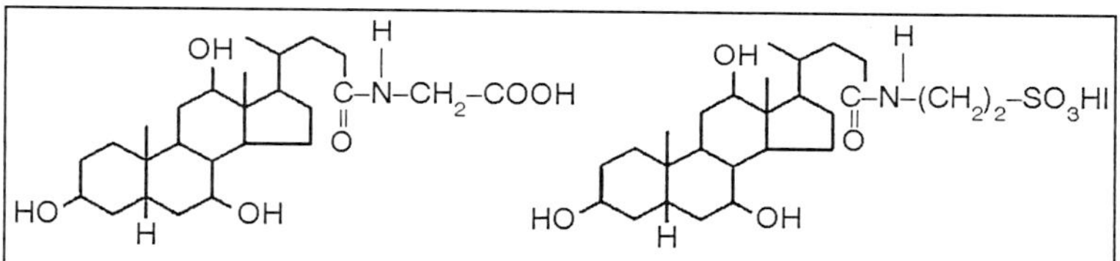

Glycochlic Acid

Taurocholic Acid

STRUCTURE OF CONJUGATED CHOLIC ACIDS

Within the liver the carboxyl group of primary and secondary bile acids is conjugated via an amide bond to either glycine or taurine before their being resecreted into the bile canaliculi. These conjugation reactions yield glycoconjugates and tauroconjugates, respectively. The bile canaliculi join with the bile ductules, which then form the bile ducts. Bile acids are carried from the liver through these ducts to the gallbladder, where they are stored for future use.The ultimate fate of bile acids is secretion into the intestine, where they aid in the emulsification of dietary lipids. In the gut the glycine and taurine residues are removed and the bile acids are either excreted (only a small Percentage) or reabsorbed by the gut and returned to the liver. This process of secretion from the liver to the gallbladder, to the intestines and finally reabsorbtion is termed the enterohepatic circulation.

CLINICAL SIGNIFICANCE OF BILE ACID SYNTHESIS

Bile acids perform four physiologically significant functions:

- Their synthesis and subsequent excretion in the feces represent the only significant mechanism for the elimination of excess cholesterol.

- Bile acids and phospholipids solubilize cholesterol in the bile, thereby preventing the precipitation of cholesterol in the gallbladder.
- They facilitate the digestion of dietary triacylglycerols by acting as emulsifying agents that render fats accessible to pancreatic lipases.
- They facilitate the intestinal absorption of fat-soluble vitamins.

DISORDER/ DISEASES OF LIPID METABOLISM

DISORDERS OF LIPID METABOLISM

Fats (lipids) are an important source of energy for the body. The body's store of fat is constantly broken down and reassembled to balance the body's energy needs with the food available. Groups of specific enzymes help the body break down and process fats. Certain abnormalities in these enzymes can lead to the buildup of specific fatty substances that normally would have been broken down by the enzymes. Over time, accumulations of these substances can be harmful to many organs of the body. Disorders caused by the accumulation of lipids are called lipidoses. Other enzyme abnormalities prevent the body from converting fats into energy normally. These abnormalities are called fatty acid oxidation disorders.

GAUCHER'S DISEASE

Gaucher's disease is caused by a buildup of glucocerebrosides in tissues. Children who have the infantile form usually die within a year, but children and adults who develop the disease later in life may survive for many years.

In Gaucher's disease, glucocerebrosides, which are a product of fat metabolism, accumulate in tissues. Gaucher's disease is the most common lipidosis. The disease is most common among Ashkenazi (Eastern European) Jews. Gaucher's disease leads to an enlarged liver and spleen and a brownish pigmentation of the skin. Accumulations of glucocerebrosides in the eyes cause yellow spots called pingueculae to appear. Accumulations in the bone marrow can cause pain and destroy bone.

RARE HEREDITARY DISORDERS OF LIPID METABOLISM

Wolman's disease results when specific types of cholesterol and glycerides accumulate in tissues. This disease causes enlargement of the spleen and liver. Calcium deposits in the adrenal glands cause them to harden, and fatty diarrhea (steatorrhea) also occurs. Infants with Wolman's disease usually die by 6 months of age.

Cerebrotendinous xanthomatosis occurs when cholestanol, a product of cholesterol metabolism, accumulates in tissues. This disease eventually leads to uncoordinated movements, dementia, cataracts, and fatty growths (xanthomas) on tendons. The disabling symptoms often appear after age 30. If started early, the drug chenodiol helps prevent progression of the disease, but it cannot undo any damage already done.In sitosterolemia, fats from fruits and vegetables accumulate in blood and tissues. The buildup of fats leads to atherosclerosis, abnormal red blood cells, and xanthomas on tendons. Treatment consists of reducing the intake of foods that are rich in plant fats, such as vegetable oils, and takingcholestyramine resin. In Refsum's disease, phytanic acid, which is a product of fat metabolism, accumulates in tissues. A buildup of phytanic acid leads to nerve and retinal damage, spastic movements, and changes in the bone and skin. Treatment involves avoiding eating green fruits and vegetables that contain chlorophyll. Plasmapheresis, in which phytanic acid is removed from the blood, may be helpful.Type 1, the chronic form of Gaucher's disease, is the most common. It results in an enlarged liver and spleen and bone abnormalities. Most commonly diagnosed during adulthood, type 1 Gaucher's disease may lead to severe liver disease, including increased risk of bleeding from the stomach and esophagus and liver cancer. Nervous system problems can also occur.

Type 2, the infantile form, usually causes death in the first year of life. Affected infants have an enlarged spleen and severe nervous system problems.Type 3, the juvenile form, can begin at any time during childhood. Children with type 3 disease have an enlarged liver and spleen, bone abnormalities, and slowly progressive nervous system problems. Children who survive to adolescence may live for many years.

Many people with Gaucher's disease can be treated with enzyme replacement therapy, in which enzymes are given by vein, usually every 2 weeks. Enzyme replacement therapy is most effective for people who do not have nervous system complications.

TAY-SACHS DISEASE

Tay-Sachs disease is caused by a buildup of gangliosides in the tissues. This disease results in early death.In Tay-Sachs disease, gangliosides, which are products of fat metabolism, accumulate in tissues. The disease is most common among families of Eastern European Jewish origin. At a very early age, children with this disease become progressively intellectually disabled and appear to have floppy muscle tone. Spasticity develops and is followed by paralysis, dementia, and blindness. These children usually die by age 3 or 4. The disease cannot be treated or cured.

Before conception, parents can find out whether they carry the gene that causes the disease. During pregnancy, Tay-Sachs disease can be identified in the fetus by chorionic villus sampling or amniocentesis.

NIEMANN-PICK DISEASE

Niemann-Pick disease is caused by a buildup of sphingomyelin or cholesterol in the tissues. This disease causes many neurologic problems. In Niemann-Pick disease, the deficiency of a specific enzyme results in the accumulation of sphingomyelin (a product of fat metabolism) or cholesterol. Niemann-Pick disease has several forms, depending on the severity of the enzyme deficiency, which determines how much sphingomyelin or cholesterol accumulates. The most severe forms tend to occur in Jewish people. The milder forms occur in all ethnic groups.

In the most severe form (type A), children fail to grow normally and have several neurologic problems. These children usually die by age 3. Children with type B disease develop fatty growths in the skin, areas of dark pigmentation, and an enlarged liver, spleen, and lymph nodes. They may be intellectually disabled. Children with type C disease develop symptoms during childhood, with seizures and neurologic deterioration. Some forms of Niemann-Pick disease can be diagnosed in the fetus by chorionic villus sampling or amniocentesis. After birth, the diagnosis can be made by a liver biopsy (removal of a tissue specimen for examination under a microscope). None of the types of Niemann-Pick disease can be cured, and children tend to die of infection or progressive dysfunction of the central nervous system. Currently, some therapies that may slow or halt the progression of symptoms in types B and C are being studied.

FABRY'S DISEASE

Fabry's disease is caused by a buildup of glycolipid in tissues. This disease causes skin growths, pain in the extremities, poor vision, recurrent episodes of fever, and kidney or heart failure.In Fabry's disease, glycolipid, which is a product of fat metabolism, accumulates in tissues. Because the defective gene for this rare disorder is carried on the X chromosome, the full-blown disease occurs only in males (Genetics: X-Linked Inheritance).

The accumulation of glycolipid causes Non-cancerous skin growths (angiokeratomas) to form on the lower part of the trunk. The corneas become cloudy, resulting in poor vision. A burning pain may develop in the arms and legs, and children may have episodes of fever. Children with Fabry's disease eventually develop kidney failure and heart disease, although most often, they live into adulthood.Kidney failure may lead to high blood pressure, which may result in stroke. Fabry's disease can be diagnosed in the fetus by chorionic villus sampling or amniocentesis. The disease cannot be cured or even treated directly, but researchers are investigating a treatment in which the deficient enzyme is replaced by transfusion. Treatment consists of taking analgesics to help relieve pain and fever or anticonvulsants. People with kidney failure may need a kidney transplant.

FATTY ACID OXIDATION DISORDERS

Fatty acid oxidation disorders are caused by a lack or deficiency of the enzymes needed to break down fats, resulting in delayed mental and physical development. Several enzymes help break down fats so that they may be turned into energy. An inherited defect or deficiency of one of these enzymes leaves the body short of energy and allows breakdown products, such as acyl-CoA, to accumulate. The enzyme most commonly deficient is medium chain acyl-CoA dehydrogenase (MCAD). Other enzyme deficiencies include short chain acyl-CoA-dehydrogenase deficiency (SCAD), long chain-3-hydroxyacyl-CoA-deficiency (LCHAD), and trifunctional protein deficiency (TFP).

MCAD Deficiency: This disorder is one of the most common inherited disorders of metabolism, particularly among people of Northern European descent.

Symptoms usually develop between birth and age 3. Children are most likely to develop symptoms if they go without food for a period of time (which depletes other sources of energy) or have an increased need for calories because of exercise or illness. The level of sugar in the blood drops significantly, causing confusion or coma. Children become weak and may have vomiting or seizures. Over the long term, children have delayed mental and physical development, an enlarged liver, heart muscle weakness, and an irregular heartbeat. Sudden death may occur.

Since 2007, nearly every state in the United States has screened all newborns for MCAD with a blood test. Immediate treatment is with glucose given by vein. For long-term treatment, children must eat often, never skip meals, and consume a diet high in carbohydrates and low in fats. Supplements of the amino acid carnitine may be helpful. The long-term outcome is generally good.

7

Genetics Behaviour

Although problems of definition and measurement are important in any area of Behavioural investigation, they are especially critical in Behaviour genetic analyses, for unless a Behavioural phenotype is consistently and unambiguously measured, there cannot be consistency and unambiguity regarding its genetic determinants or its relationship to other Behavioural, somatic, or morphological phenotypes. There are few areas of Behavioural investigation in which this "phenotype problem" is so clearly shown as in the genetic analysis of activity and exploratory Behaviour directly and indirectly.

Activity and exploratory Behaviour are fundamental, yet surprisingly complex Behavioural phenotypes. Important in their own right for the organism, activity and exploration are also of importance to researchers in that each of these Behaviours has been used as an index for more abstract Behavioural concepts, such as emotionality and curiosity. We have dealt solely with the genetic determinants of activity and exploratory Behaviour.

To do so adequately we needed to go into some detail regarding definitional problems and techniques of measurement. We did not intend to provide a thorough review of the voluminous literature on these topics, but rather have emphasized studies that provide clear statements of genetic effects or of a genetic model, or that serve as definitive examples of a particular technique. An additional criterion in the selection of a major reference was to select, where possible, the study that would be most accessible to the greatest number of readers.

DEFINITION

Any movement of any part of the body of an animal is, according to a broad definition of the term, *activity*. However, as we use the term, and as it is most commonly used in the literature, "activity" refers to movement of the whole organism.In order to distinguish whole organism movement from such Behaviours as orienting response (head turning responses, movement of the pinnae, or movement of the eyes), the term *locomotor activity* is often used. Fuller and Thompson recommend *ambulation* as a more unambiguous term. The problems with activity as a phenotype, however, derive not so much from nomenclature as from what this form of Behaviour represents, and how it is measured.

Activity as a Trait

For several decades prior to the 1960s, activity *qua* activity was outside the mainstream of American psychology. During this era of the major Behaviouristic learning theories, the concept of "Behavioural homeostasis" was in vogue. This concept held that the "normal" state of an organism was not behaving (*i.e.*, being inactive), and that activity occurred only as the result of a "drive," that resulted from an imbalance in a physiological system, such as a tissue deficit or increase in hormonal activity.

According to the assumption of Behavioural homeostasis, then, one does not (indeed, cannot) account for activity as a general trait, because activity occurs only as a consequence of other internal and external events. In recent years, the waning influence of the major learning theories, together with a number of disparate empirical findings (for example, exploration of their surroundings by satiated animals; the neural and bodily activity seen in many species during sleep), has led to a decline in the acceptance of the Behavioural homeostasis assumption. There is an increasing degree of acceptance of its opposite: The normal state of an organism is behaving (*i.e.*, being active).

When comparing the amount of movement of different individuals within similar settings, in the absence of specific physiological states, external goals, or noxious stimuli, one is investigating activity as a trait.

If individual organisms do indeed possess specific "general activity levels" that yield strong predictions of other sorts of Behaviour or that are highly consistent across environmental settings and resistant to alteration, the theoretical and practical implications would indeed be great. Knowledge of the extent and nature of the genetic determinants of activity as a trait would magnify the importance of these implications.

ACTIVITY AS AN INDEX

Activity has been of interest not only as a trait in its own right, but also as an index, or means of measurement, of more complex and difficult to define forms of Behaviour. The most frequent uses for activity measures in this regard are exploratory Behaviour, whiich we discuss at length and emotionality. As an index of emotionality, activity is often combined with other measures, especially urination and defecation. Stemming largely from the selection studies in rats by Hall in the early 1930s, emotionality is, in turn, an indication of the autonomic and Behavioural responsiveness of organisms to potentially dangerous or noxious situations. Organisms are assumed to be either more active or much less active (*e.g.*, "freezing") in these situations, depending on the species or the genotype within a species.

The advantages of using activity as an index are obvious:

- Activity is relatively easy to measure, with a number of measurement procedures yielding reliable, readily interpretable scores
- A considerable body of activity data exist for a variety of species of labouratory animals
- A number of successful selection and strain comparison studies have been completed, providing an abundance of subjects more or less "genetically standardized" on this phenotype.

The problems associated with using activity as an index are more subtle, but no less real. If activity serves as an index of a more complex characteristic, such as emotionality, there should be a strong correlation between activity and any other index measures. Such a correlation would indicate that the various phenotypes bear some causal relationship to one another; or that they are due to pleiotropic or linkage effects of the same genes; or, on the other hand, that the correlation between indices is a spurious, chance occurrence, perhaps furthered by measurement artifacts.

When the correlation between indices is high for some selected lines but not for others, or for some inbred strains but not others, one must give the greatest credence to the last of these explanations. We deal with this problem, a major factor in phenotype confusability.When activity is used as the sole index of a more complex phenotype, it is the same as operationally defining the more complex phenotype in terms of the measure of activity (for example, defining exploratory Behaviour *as* "squares crossed in an open field"). It then becomes the obligation of the "consumers" of such information to decide for themselves whether this sort of measure is appropriately definitive and adequately representative of the complex phenotype.

An example of the problems that can arise from the use of activity as an index can be illustrated by a study performed by Whimbey and Denenberg, and further discussed by Denenberg in 1969. In the portions of this study relevant to our discussion, they placed rats in an open field briefly for 5 days and obtained a variety of measures that were then intercorrelated and factor analysed.

Correlated positively with defecation scores and could be considered an index of "emotionality." However, on Day 2 and thereafter this same measure of activity, taken under presumably the same conditions from the same subjects, produced a *negative* correlation with defecation and yielded different factor loadings. Then, the authors imply, activity becomes an index of exploratory Behaviour. Is activity a useful index for more complex Behavioural phenotypes?

MEASUREMENT

THE ACTIVITY WHEEL

This simple, time honoured device is still available from a number of labouratory equipment suppliers in sizes suitable for rats, mice, and other small animals. These devices usually operate a counter that provides an easily understood and interpreted measure of activity expressed as the number of revolutions per unit of time. (Most commercial activity wheels will run both clockwise and counterclockwise, and will operate the counter when turned in either direction.) In addition to obvous extraneous variables such as room temperature, illumination level, and the like, the major variables to be controlled with this type of apparatus are:

- The degree of effort needed from the subject to overcome friction in order to turn the wheel; and
- The degree of choice available to the animal to move freely from an adjoining cage into the wheel. Assuming the greatest possible control, the activity wheel provides what is, perhaps, the purest measure of ambulation, at least in terms of energy output. However, the very conditions providing the control and "purity" of this measurement technique also make it the most artificial; ordinarily, the movement of an organism results in a changing environment for the animal as it progresses. Inasmuch as running in the wheel does not lead to changes in the animal's environment and does not therefore result in exploratory or neophobic excursions, it is not surprising that the "distance" run in such a device is often greater than the distance traversed under more natural conditions. Perhaps this type of activity, unlike that measured by other means, is primarily exercise.

Maze Running

Where the activity wheel allows only totally circumscribed locomotion, the maze allows some degree of directionality of movement. The maze thus measures ambulation in a relatively more naturalistic fashion than does the activity wheel, especially for thigmotaxic and burrowing species, such as rats. However, it is possible that an animal moving through an unfamiliar maze is also exploring its environment, or seeking an escape route, or is engaged in some other form of Behaviour that involves ambulation. Therefore, attempting to measure *activity* in a more realistic manner can create questions as to just what is being measured.

Activity Metres

These are automated measuring devices that unobtrusively detect and count movements of animals. Newer versions consist of a platform containing a series of electromagnetic sensors upon which any Non-metallic container with one or more animals inside can be placed. Any movement by the animal operates a counter, with the degree of movement required for detection controllable by setting the sensitivity of the sensors. The activity metre has the unique property of permitting measurement of movement in any sort of box or container that will fit on its surface, including home cages. Another unique property of this type of device is that it measures all movement by the subject, rather than merely ambulation from one location to another.Although this has the advantage of being a meaningful measure of activity as a trait, it dictates caution when comparing scores with those obtained in other types of activity measuring devices. The unique and specific conditions that must be controlled when using an activity metre involve the placement of the animal container over the sensors and the sensitivity setting of the metre.

Although the raw scores taken from the counter activated by movement over the sensors are difficult to interpret, meaningful and statistically reliable ordinal data have been obtained with this device in studies comparing genotypes and experimental conditions.

The Open Field

This is the most frequently used technique for the measurement of activity, a fact that should have resulted in a major advantage for its use: the establishment of reliable standardized scores to serve as a basis for comparison across genotypes and experimental conditions. There is also a high degree of face validity for its use in the measurement of activity as a trait—if ambulation within an open arena is not activity, what is?

The open field can be constructed in a variety of shapes (round, square, and rectangular being the most popular) and made in sizes appropriate for almost any species. More limited versions of this apparatus are the straight runway and the shuttle box, which share many of the advantages and disadvantages found with the open field.

Unfortunately, despite its apparent advantages, the open field test has not turned out to be free of severe problems. As Walsh and Cummins point out in their thorough and critical review of the many studies using this technique: "Apparatus, techniques, subjects, parametres, analyses, and interpretations have diversified enormously, yet in the literature, sweeping generalizations and conflicting interpretations continue to be made on the basis of univariate studies lists the variety of Behavioural and autonomic measures that have been taken, although number of squares crossed seems to be the most common measure in most Behaviour genetic studies. One of the difficulties referred to by Walsh and Cummins results from the attempt to compare studies that have used different measures.

GENETIC DIFFERENCES

The presence of a genetic effect was originally established for activity through selective breeding of animals showing high activity scores, low activity scores, or both on a standardized measuring technique. One of the earliest attempts to establish the hereditary basis of activity using this method was a selection study by Rundquist, in which he selected rats for high and low activity following testing in an activity wheel. His results were disappointing initially, because of an error in the method of selection he employed: the mating of high active and low active offspring from Non-selected parents. That is, rats showing high activity scores were selected not only from parents that also had high activity scores, but from those yielding low activity scores as well, with the reverse holding true for rats with low activity scores. As a result, there was no differentiation between groups from generations F_1 to F_4. When this error was corrected, differentiation into high and low active groups did occur and laid the foundation for further research in this area.

Hall performed selection experiments in the early 1930s to establish the genetic basis of emotionality in rats using as phenotypes the Behaviour of the rats in an open field. Although he was not concerned with activity per se, his selection procedures and his use of the open field test have served as the prototype for many of the modem studies on this problem.

In the 1950s, it became possible to use another method for determining hereditary basis for activity and for any other phenotype. This method involves finding differences in activity between strains of genetically controlled animals, usually highly inbred strains of mice and rats, but also established breeds of dogs and inbred strains of various species of *Drosophila.*

Thompson tested five strains of inbred mice on two activity measures, the open field and the Y maze. His results showed marked differences in the activity levels of the five strains, and the two C_{57} strains showed much higher activity levels in both open field test and the Y maze than did the BALB/c mice.

In mouse studies, McClearn has shown that there is a clear relationship between locomotor activity and the per centage of C_{57}BL/Crgl genes present (from backcrosses with A/Jax mice). Thompson illustrated that there is considerable variability between strains in exploratory activity, with C_{57}BR/a, C_{57}BL/6, and C_{57}BL/10 having the highest mean score (for activity) and A, AK/e, and BALB/c strains having the lowest.

These orderings were fairly invariant with another testing apparatus (enclosed arena versus Y maze). There are differences in the ordering of mice on exploratory activity with different environmental conditions. For example, he found that in bright white light, A/Crgl mice showed less exploratory activity than in dim red light. In the C_{57}BL/Crgl mice, activity increased in white light and decreased in dim red light. Although he found no change in the ordering of the strains, McClearn did find a significant interaction between strain and illumination.Of particular historic and current interest are the MR and MNR rats originated by Broadhurst. Using a procedure essentially similar to that used by Hall more than two decades earlier for selection for emotionality, Broadhurst made several important improvements; the open field was more uniformly and brightly lighted and initial responses of the subjects were weighted more heavily, giving less emphasis to the possible adaptation of the rats to a familiar environment. MR and MNR rats clearly differ on open field defecation scores, an index of emotionality, and are often regarded as "benchmark" strains for this category of Behaviour. However, on several *activity* measures, these strains do not differ greatly. Selection studies illustrating genetic differences in activity have also been carried out in species other than rats and mice.

The comparison of the normal Behaviour of different established breeds of dogs provides an excellent example; for "everyday" differences in activity levels, one need only compare a bassett hound and a fox terrier. Dogs have also been used as subjects in more systematic genetic studies, with different breeds treated analogously to inbred strains of rodents.

A prime example of this sort of research is the 19 year study conducted by Scott and Fuller that dealt with, among other things, the emotional Behaviour of several breeds of dogs, together with their hybrids and backcrosses. Clear genetic differences were found in the five breeds on a variety of reactivity tests, including open field Behaviour.

Some work on activity and its implications has also been done with the fruit fly. In one study flies were selected for mating speed. Activity differences for these two selected strains (fast mating speed and slow mating speed) were measured by counting the number of squares a fly entered in an open field arena over a given period of time. It was found that slow mating lines exhibited more open field activity than did fast mating lines.

In contrast, Ewing, in selecting for spontaneous activity, found that the less active lines displayed *more* sexual Behaviour than did the more active lines. Although these results appear to contradict Manning's findings, the discrepancy appears to have been due to an "apparatus effect," as he used an apparatus other than the open field used by Manning. When Ewing replicated his study using the type of open field apparatus used by Manning, no significant differences between the selected lines were found.

Besides being alterable by apparatus effects and other environmental changes, activity is readily changed, often differentially for different genotypes, by internal events, especially pharmacological agents. A thorough review of drug effects on activity, and the genetic factors involved, has recently been prepared by Broadhurst. The studies we have cited are typical of many others to be found in the literature, in that many of the findings are contradictory with one another, with varying differences among genotypes (whether inbred strains or selected lines) depending on procedural details, types of apparatus, and so on. Therefore, generalizations on activity levels of specific genetically controlled animals should be treated cautiously.

An excellent example illustrating this point can be found in an article by Lassalle and Le Pape. They demonstrated that the environmental conditions associated with testing of two inbred mouse strains (C_{57}BL/6 and BALB/c) can reverse the general finding we referred to earlier that C_{57}s are more active than BALB/cs. Under seminatural and breeding cage conditions, the BALB/cs showed a higher level of activity than did the C_{57}s, and in the seminatural condition, the BALB/cs were more active in the dark).

In an issue related to the genetic determination of activity, several studies have attempted to determine heritabilities and degrees of genetic determination. However, as these findings are limited to the specific genetic population measured within a specific type of apparatus and a specific set of environmental conditions, these data are of rather limited use in the generalizable information they provide for an understanding of the genetic basis of activity.

MODELS

There are various avenues of research that have suggested that a polygenic model can best account for activity. One form of evidence for this type of genetic determination can be found in score distributions; typically, a continuous, unimodal distribution of scores has been found for most types of activity measurement when genetically heterogeneous animals were tested.

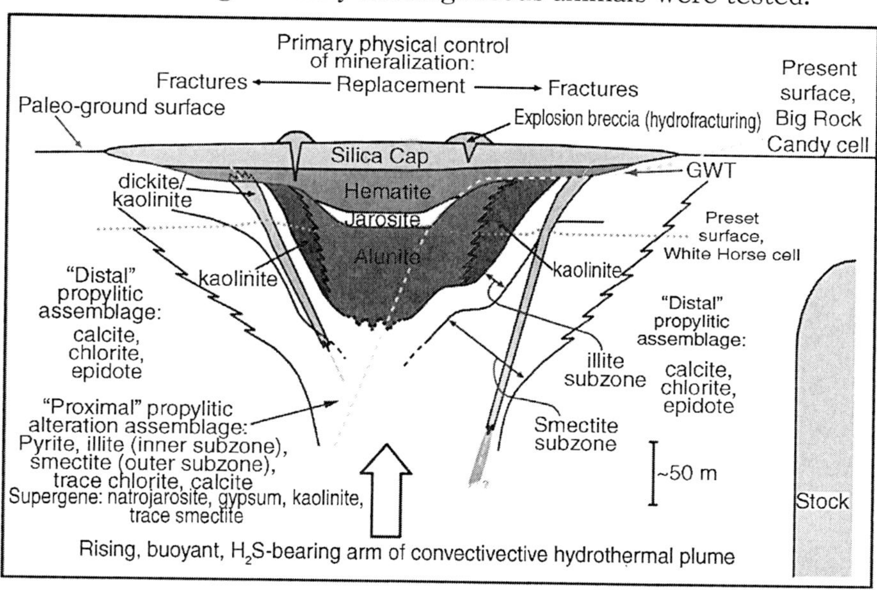

Fig. Genetic Models

Tryon suggests a polygenic model on the basis of this form of evidence alone. An excellent example of a more thorough analysis supporting a polygenic model can be found in the series of studies reported by DeFries and Hegmann. They tested large numbers of two inbred strains of mice (C_{57}BL/6J and BALB/cJ), as well as their reciprocal F_1 and F_2 hybrids and backcrosses.

The mice were tested for 3 minutes on each of 2 successive days in an automated, white Plexiglas open field, 90 cm^2. The means and variances of the transformed scores. The degree of increase in variance from the F_1 to the F_2 generation is inversely related to the number of loci affecting a trait. The variance between the F_1 and F_2 generations is very slight, especially when compared with the difference in variances between the parental strains and the F_1 generation. (One can discount the variance differences between the reciprocal F_2 s as a maternal effect, especially as it diminishes in later generations.) DeFries and Hegmann conclude that activity is controlled by genes at several loci, suggesting *at least* 3.2 loci from an earlier analysis by Hegmann.

A "mineral occurrence" is a concentration of a mineral (usually, but not necessarily, considered in terms of some commodity, such as copper, barite or gold) that is considered valuable by someone somewhere, or that is of scientific or technical interest. In rare instances (such as titanium in a rutile bearing black sand), the commodity might not even be concentrated above its average crustal abundance. A "mineral deposit" is a mineral occurrence of sufficient size and grade that it might, under the most favourable of circumstances, be considered to have economic potential.An "ore deposit" is a mineral deposit that has been tested and is known to be of sufficient size, grade, and accessibility to be producible to yield a profit. (In these days of controlled economies and integrated industries, the "profit" decision may be based on considerations that extend far beyond the mine itself, in some instances relating to the overall health of a national economy.)

On one hand, the field observations usually begin with "mineral occurrences" (or with clues to their existence) and progress with further study to "mineral deposits" and only rarely to "ore deposits," but we must present information that helps us deal with all classes of "mineral occurrences," not just "ore deposits." On the other hand, in terms of accessible information our sample is strongly biased Towards "ore deposits," for it is only in them that sufficient exposure is available to develop a real knowledge of the overall character of the mineralization process. Some mineral occurrences are, therefore, unrecognized mineral deposits, while others are simply mineralized localities where ore forming processes were so weak or incomplete that a deposit was not formed. Thus we summarize the state of knowledge regarding ore deposit models, and we call them "mineral deposit models" with the hope that what we have learned about large and high grade metal concentrations will help us sort out all mineral occurrences to identify their true character and, we hope, to recognize which have potential to constitute ore deposits.

The attributes or properties of a mineral occurrence are, of course, those features exhibited by the occurrence. When applied to a model, these terms refer to those features possessed by the class of deposits represented by the model. It is useful to consider attributes on at least two scales: the first deals with local features that may be observed directly in the field (mineralogy, zonal patterns, local chemical haloes, and so on); the second is those features concerning the regional geologic setting and which must be interpreted from the local studies or may be inferred from global tectonic considerations (for instance, that the rock sequence under study represents a deep water, back arc rift environment, or that the area is underlain by anomalously radioactive high silica rhyolite and granite). Two of the most prominent attributes, the commodities/geochemical patterns and the mineralogy, are cross indexed to model types.

To the greatest extent possible, models were constructed so as to be independent of site specific attributes and therefore contain only those features which are transferable from one deposit to another. This goal is difficult to attain, because we do not always know which features are site specific.The term "model" in an earth science context elicits a wide variety of mental images, ranging from the physical duplication of the form of a subject, as in a scale model of the workings of a mine, to a unifying concept that explains or describes a complex phenomenon. In this context we shall apply only the latter usage. Therefore, let us propose a working definition of "model" in the context of mineral deposits, the overriding purpose being to communicate information that helps mankind find and evaluate mineral deposits. A mineral deposit model is the systematically arranged information describing the essential attributes (properties) of a class of mineral deposits. The model may be empirical (descriptive), in which instance the various attributes are recognized as essential even though their relationships are unknown; or it may be theoretical (genetic), in which instance the attributes are interrelated through some fundamental concept.One factor favouring the genetic model over the simply descriptive is the sheer volume of descriptive information needed to represent the many features of complex deposits. If all such information were to be included, the number of models would escalate until it approached the total number of individual deposits considered.

Thus we should no longer have models, but simply descriptions of individual deposits. Therefore, the compilers must use whatever sophisticated or rudimentary genetic concepts are at their disposal to distinguish the critical from the incidental attributes. It is commonly necessary to carry some possibly superficial attributes in order not to preclude some permissible but not necessarily favoured, multiple working concepts.

One of the commonly accepted attributes of the model for the carbonate hosted lead zinc deposits of the Mississippi Valley type is the presence of secondary dolomite. But do we know that this is essential? Suppose a deposit were found in limestone; would we reject its assignment to the Mississippi Valley class? Or could it be correct that the critical property is permeability and that the formation of dolomite either:

- Enhances permeability (and thereby makes the ground more favourable),
- Reflects pre existing permeability that is exploited by both the dolomite and the ore?

MODEL NAMES

Each model has been assigned a name that is derived either from the special characteristics of the classes or from a type locality. The latter strategy was employed to avoid excessively long descriptive names. The use of type names derived from specific deposits does produce confusion in some readers, however, who may feel, for example, that a deposit that does not look "exactly" like Comstock cannot be represented by a "Comstock epithermal vein" model. This confusion may be minimized by realizing that most models are blends of attributes from a large number of deposits and that the names are only conveniences, not constrictions. The contributors to this report and the literature in general are not without disagreements regarding nomenclature (as well as genetic aspects and some facets of the groupings made here), but provision for alternative names is made in the model format under the heading of approximate synonyms.

DESCRIPTIVE MODELS

Because every mineral deposit, like every fingerprint, is different from every other in some finite way, models have to progress beyond the purely descriptive in order to represent more than single deposits.

Deposits sharing a relatively wide variety and large number of attributes come to be characterized as a "type," and a model representing that type can evolve. Generally accepted genetic interpretations play a significant role in establishing model classes. Here we shall emphasize the more descriptive aspects of the deposits because our goal is to provide a basis for interpreting geologic observations rather than to provide interpretations in search of examples. The attributes listed are intended to be guides for resource assessment and for exploration, both in the planning stage and in the interpretation of findings.

The descriptive models have two parts. The first, the "Geological Environment," describes the environments in which the deposits are found; the second gives the identifying characteristics of the deposits. The headings "Rock Types" and "Textures" cover the favourable host rocks of deposits as well as source rocks believed to be responsible for hydrothermal fluids which may have introduced epigenetic deposits. "Age" refers to the age of the event responsible for the formation of the deposit. "Tectonic Setting" is concerned with major features or provinces (perhaps those that might be portrayed only at 1:1,000,000 or smaller scale), not ore control by structures that are local and often site specific. "Associated Deposits" are listed as deposits whose presence might indicate suitable conditions for additional deposits of the type portrayed by the model. The second part of the model, the "Deposit Description," provides the identifying characteristics of the deposits themselves, particularly emphasizing aspects by which the deposits might be recognized through their geochemical and geophysical anomalies.

In most cases the descriptions also contain data useful in project planning for mineral assessment or exploration; this aspect is especially important where limited financial and manpower resources must be allocated to the more significant tasks.

GRADE TONNAGE MODELS

Estimated premining tonnages and grades from over 3,900 well explored, well characterized deposits were used to construct 60 grade tonnage models. Where several different estimates were available for a deposit, the estimated tonnages associated with the lowest cutoff grades were used. Grades not available (always for by products) were treated as zero. Except for a few instances, the data base is so large as to preclude specific references. Several compilations of data were particularly useful sources for multiple deposit. The grade tonnage models are presented in graphical format to make it easy to compare deposit types and to display the data.

All plots show either grade or tonnage on the horizontal axis, while the vertical axis is always the cumulative proportion of deposits. Plots of the same commodity or tonnages are presented on the same scale; a logarithmic scale is used for tonnage and most grades. Each dot represents an individual deposit (or, rarely, a district), cumulated in ascending grade or tonnage.

Where a large number of deposits is plotted, individual digits represent the number of deposits. Smoothed curves are plotted through arrays of points, and intercepts for the 90th, 50th, and 10th per centiles are constructed. For tonnages and most grades, the smoothed curves represent per centiles of a lognormal distribution that has the same mean and standard deviation as the observed data; exceptions are plots where only a small per centage of deposits had reported grades and grade plots that are presented on an arithmetic scale, such as iron or manganese, for which the smoothed curve was fit by eye.

The number of deposits in each type is indicated at the upper right of each diagram. The deposits used to construct each model are listed with the model and cross indexed to model types. Correlations among grades and between tonnage and each grade are indicated only when significant at the 1 per cent level.There are important limitations inherent in the data base used for all grade tonnage models. Estimates of cutoff grades within individual deposit types can vary because of regional, national, or operator differences.All too commonly there is no mention of the actual cutoff grades or mining widths that are incorporated; nevertheless, the grade tonnage figures given do represent material that the company or the government believed might someday be economic to mine. Stratiform deposits of large areal extent, such as phosphate or sedimentary manganese, are special problems because of differences in opinion and practice regarding how closely drilled they must be to "prove" ore tonnages and regarding the thicknesses and depths of what may be considered for eventual mining.

Effects of another source of variation, mining methods, are recognized in some of the placer models; typically, however, mining methods are fairly consistent within a deposit type. In a few instances, irregular cumulative frequency plots reflect mixing of economic and scientific data sources, such as in the plot of gold in porphyry copper deposits. In spite of the current difficulty of quantifying variation of grades and tonnages with respect to changes in cutoff grades or mining methods, the models presented here are believed to account for the main source of variation in grades and tonnages of mineral deposits variation due to differences among types of deposits.

The question of whether one counts deposits within a cluster of related deposits as individuals or as a total will probably never be resolved to everyone's satisfaction. Some geostatisticians would separate each ore body (and then argue about whether two operations on the same body should be counted separately), whereas some economic geologists would lump everything from a single district (and then argue about district boundaries). For the most part the entities summarized are individual deposits, but in some instances such data are mixed with data representing entire districts. Because of these inconsistencies, some care is necessary in comparing grade tonnage models between deposit types or in comparing this summary with those prepared using alternative methods.

Care is also warranted in interpreting the grade distributions for which data are missing; this concerns principally by product grades. In some instances, such as the platinum group element (PGE) contents in podiform chromite and the cobalt content of laterites, the fragmentary information given probably represents the entire class.

In other instances, such as the lead content of Cyprus massive sulfide deposits, the missing grades probably represent values below the lowest reported grades. The grades derived from studies of trace elements in ores more probably represent the former situation rather than the latter. Deposits strongly suspected to be small or very low grade are seldom sampled well enough to be characterized in terms of grade and tonnage, thus the sample of many deposit classes is truncated by economics.

Non-etheless, probably 40 per cent of the deposits used in these models are, in fact, non economic today; and a perusal of the figures will discover examples of both small deposits and low grade deposits. Potential metal supply is dominated by the very few largest tonnage deposits, as shown by Singer and

DeYoung (1980), who also pointed out that inverse correlations between grade and tonnage are surprisingly rare. Thus the fact that a deposit is large does not necessarily mean that it will prove to be of low grade. This means that most low grade deposits are not likely to have huge resources and also that the omission of a few low grade or small tonnage deposits will not seriously degrade the predictions of potential national supplies for most commodities.

In contrast, the missing low grade and small deposits suggest that the grade tonnage models represent a biased sample of the large number of low grade or small tonnage occurrences and prospects found by exploration.

This fact must be considered in cases where the number of undiscovered deposits is estimated. In order for the estimated number of deposits to be consistent with a grade tonnage model, approximately half of the deposits estimated should have greater than the model's median tonnage or grade. Thus the probability that an untested prospect represents a significant deposit can too easily be overestimated.

OTHER TYPES OF MODELS

The bulk of this report deals with descriptive mineral deposit models and their grade tonnage counterparts, but there are other useful aspects which we wish to discuss even though we have not yet had the opportunity to develop or exploit them. They are the genetic, occurrence probability, and quantitative process models.

Many authors prefer to keep a clear distinction between descriptive and genetic models, apparently feeling that the descriptive models somehow represent "pure truth" whereas the genetic constitute a less objective philosophical position (or at least make the investigator "skate on thin ice"). It is altogether desirable to avoid confusing interpretation with fact; but it is well to remember, for example, that each time a field geoscientist extrapolates geology across a covered area he or she adds an element of "interpretation" to a "factual" map, and that this interpretation is not necessarily any more "real" (or "unreal") than, for example, an isotope geologist's conclusion that a given oxygen and hydrogen isotopic signature extracted from fluid inclusions points to a meteoric origin for the fluid.

The point is that the whole of our professional knowledge rests on a broad continuum of interpretations; many of them are so commonly accepted that they are no longer questioned, but many others still evoke challenges. Thus we suggest that a combination descriptive genetic model is not inconsistent with professional practice. The model begins as a description, but various aspects of the model become genetic as they acquire satisfactory genetic explanations. Eventually much of the model becomes genetic, as has happened, for example, with the Cyprus type massive sulfide deposits or the sandstone uranium deposits of the Colourado Plateau.

As the attributes of a model become understood in a genetic sense, the descriptive model evolves to a genetic model:

- F models are compilations of the properties of a group of related deposits in which the reasons for certain attributes being favourable are identified. Descriptive models evolve into genetic models, and as such they become far more flexible and powerful.

  We have presented the three model subtypes above as if they constituted a linear logical sequence leading Towards the "final" model, but in fact there must be an iterative relationship among descriptive, genetic, and grade/tonnage models. The consequence of examining any of these three may be a reassessment of the groupings of deposits chosen to be represented by a model type and the redesignation of the attributes diagnostic for that type.

  With a dominantly genetic model in hand, two more model types can be generated:
- Occurrence probability models are models that predict the probability of a deposit (of a size and grade indicated by the appropriate grade tonnage models) occurring within a given area. As with the descriptive and genetic models, probability models that are tied to lithic or structural geologic entities (that is, they are genetic) are far more focused; in fact, it is probably impossible to generate a useful probability model before the establishment of a genetic model. Accurate probability models are very difficult to construct because although the technical community has very complete data on mineral producers (mines), the data on Non- producing mineral deposits (prospects and mineral shows) are much less well documented, a point also covered in the discussion of grade tonnage models. Even more importantly, data on barren areas are sparse.

  We must extrapolate from a very fragmentary base Towards a completely unseen target.

  There is much to learn before the probability model can be made a dependable tool; yet the successful targeting of exploration programmes by industry demonstrates that, at least on a qualitative basis, areas with better than average probabilities can be identified. It is worth noting, also, that mineral fuels are much more predictable and now can have realistic probability of occurrence values attached to specific volumes of sediments provided that the initial character and post depositional histories of the sediments are well known. It is a distant but not unreasonable dream to anticipate that some day we shall approach that level of certainty for some types of Non-fuel mineral deposits.
- Quantitative process models are models that describe quantitatively

some process related to mineral deposit formation; they are offshoots of the genetic model. Examples would be models of heat or fluid flow around a cooling pluton; rates of crystal growth as functions of supersaturation, impurities, and temperature; or sequences and amounts of minerals deposited from evapourating seawater.

All five of these model subtypes can be parts of the "final" model, and recycling of the model back to the original groupings stage helps refine the selection process. The persons engaged in research guidance and especially exploration and development have broad ranging needs, whereas those dealing with the availability of minerals or of land use allocation have less use for genetic or quantitative process models. Overall there is a need for a comprehensive array of mineral deposit models to meet these individual

MATURITY OF DESCRIPTIVE GENETIC MODELS

The rate at which we gain understanding and the current levels of genetic knowledge vary considerably from one deposit type to another. Such types as placers and evapourites are well known genetically and the problems in their exploration and utilization concern local site specific geologic issues rather than mineral genesis or the degree of maturation of the model.In contrast, others such as the Coeur d'Alene Ag-Pb-Zn veins, or the massive Zn-Mn-Fe oxide/silicate bodies at Franklin and Sterling Hill, or the Cu-U-Au at Olympic Dam, or the Cu-Zn-Pb-Ge ores of Kipushi and the Tsumeb pipe remain genetic enigmas despite, in the instances of the first two, extensive research spanning many years.

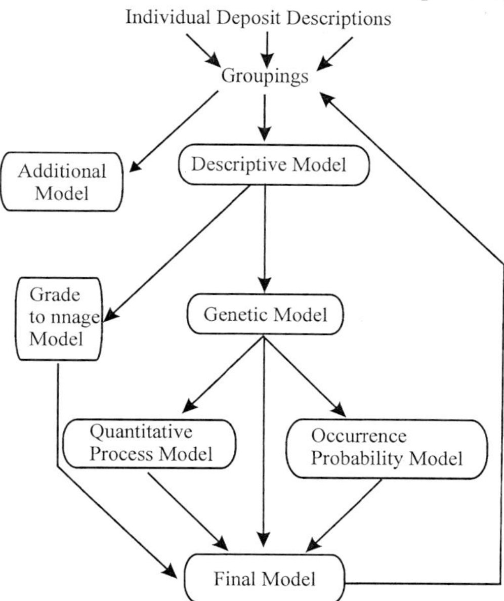

Fig. Flow Sheet Showing Evolution of Model Types.

Still others, such as the diamond bearing kimberlite pipes, are geologically well understood regarding their origin yet very poorly understood in terms of the reasons for their existing at any particular site. Our rate of acquisition of information is very irregular. The several scarps between plateaus in the knowledge curve for the marine phosphate model might mark, successively, the recognition that the phosphate was a chemical precipitate, that it occurred on continental shelves where upwelling of deep marine waters occurred, and that the upwelling regions were related to wind and current patterns that were tied to the global configuration of the continents and ocean basins.

A second example from the Mississippi Valley type ores might involve scarps marking the recognition (from fluid inclusion evidence) that the ores were deposited from warm (about 100°C) highly saline solutions that could represent neither simple surface nor marine waters. A second scarp might be associated with the recognition that the deposits were integral parts of a regional hydrologic regime whose distribution and character was susceptible to interpretation.

Bring out another point: some aspects of any model always remain to be determined, thus we never acquire a "complete" model. Indeed, the approach to "complete" understanding is asymptotic, and a lot of additional effort to clear up the "last" uncertainty in a nearly perfect model is probably unwarranted.But, new ideas and new technologies can provide the impetus for new spurts in knowledge for heretofore incomplete models. Note that the horizontal axis is simply "years of effort" devoted to fundamental geologic investigation.

The scale certainly needs to be exponential in order to fit the intensively studied and sparsely studied deposit types, but this figure is strictly schematic, there being no source of documentation for either coordinate.

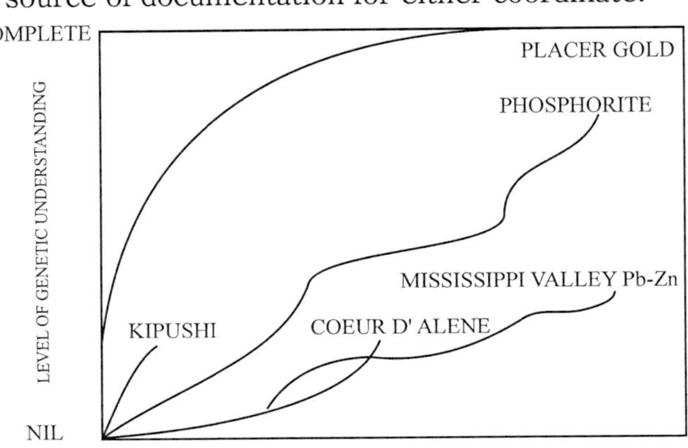

Fig. Growth Patterns of some Typical Genetic Models

The figure also indicates that different deposit types may require different amounts of effort to achieve a similar level of genetic understanding. A hypothetical growth curve along which different types of deposits have been schematically arrayed.Because some deposits (such as volcanogenic massive sulfides) are so much more difficult to understand than others (gold placers), the horizontal axis has been "normalized" by plotting a ratio of effort done to effort needed thereby permitting a smooth, although admittedly subjective and schematic, curve. There is no documentation to support this diagram, although the general concept meets with agreement among most contributors to this volume.DeFries and Hegmann report on an activity selection study they completed. As a foundation population, they used the F_3 generation from the C_{57}BL/6J and BALB/cJ mice used in the study reported earlier. Using an open field activity measure similar to that used in their earlier study, they were able to obtain two highly active, two low active, and two control lines, with each selection (and its replication) being successful within six generations.

Of particular interest for a genetic model for activity, however, was their finding that albino rats were less active than those with any other coat colouring (agouti, cinnamon, black, and brown). These data together with their findings that the albino BALB/c mice are less active than the pigmented C_{57}BL/6 mice, led DeFries and Hegmann to suggest that: "Although open field Behaviour is clearly influenced by genes at many loci, a major gene effect was found: albino Ss have lower activity and higher defecation scores than pigmented animals."

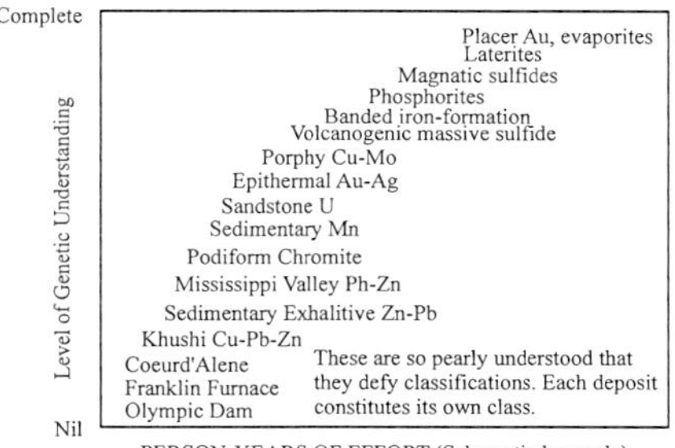

Fig. Comparison of Relative Levels of Understanding of some Important Model Types

Table. Open Field Activity Scores of C_{57}BL/6J, BALB/cJ and F_1 and F_2 Hybrid Mice

		Means [b]		*Variances*	
Generation	*Genotype* [a]	*Males*	*Females*	*Males*	*Females*
Parental	C	4.67	4.25	6.33	4.27
	B6	15.73	16.40	4.86	6.56
F_1	B6C	15.09	15.04	6.70	10.37
	CB6	13.98	13.66	15.80	21.06
F_2	CB6 x CB6	11.88	12.63	17.96	9.23
	CB6 x B6C	11.76	12.30	16.56	13.68
	B6C x CB6	12.66	12.86	15.87	16.44
	B6C x B6C	11.58	11.84	14.22	16.28

Table. Mean and Variance of Activity and Defecation Scores of Albino and Pigmented Animals in Generations $F_2 - F_5$, Pooled Across Generations

	Activity		*Defecation*		
	Mean ± S.E.	*Variance*	*Mean ± S.E.*	*Variance*	*N*
Pigmented	12.72 ± 0.18	18.65	1.78 ± 0.03	0.55	1729
Albino	10.44 ± 0.10	17.71	2.13 ± 0.02	0.49	557

This is plausible from the data presented, although it is possible that the effect found may be due to genes linked to the c locus in certain strains, with activity not actually moderated by the c locus itself. The finding that some albino (cc) mice do not exhibit a reduction in activity warrants caution in making generalizations about the "effects" of albinism. It has been found, for example, that albino LG/J mice are highly active and that albino BALB/cBy mice do not show reduced activity when tested on an activity metre. DeFries and Hegmann used BALB/cJ mice, as did several of the other studies supporting the "albino effect."

Perhaps there is linkage of the c locus with other genes that reduces activity in this strain. There could also be a different sort of linkage with the reverse effect in highly active albino strains. At this point, however, attributing reduced activity to the lack of pigmentation seems premature.

EXPLORATORY BEHAVIOUR

PHENOTYPE PROBLEM

Exploratory Behaviour is frequently confounded with, measured by, or operationally defined as ambulation (*i.e.*, activity), most often in an open field. Defining and then measuring exploratory Behaviour in this fashion creates obvious problems for any attempt to determine the nature of the genetic factors underlying exploration.

These problems are especially severe for the determination of specific genetic models. If one accepts the polygenic systems most often found for activity, and if one accepts the definition of exploratory Behaviour as ambulation within an open field (as many do), then the same polygenic models established as underlying activity also hold true for exploratory Behaviour. Following this rationale, the phenotypes—activity and exploratory Behaviour—are logically and empirically identical.

If activity and exploratory Behaviour have been shown to be identical—merely two terms for the same phenotype. Clearly, howevever, this is not the case. Although rather similar Behavioural measures and techniques have been used, the nature of the results obtained and especially the genetic models that these results have suggested point to the conclusion that activity and exploratory Behaviour are distinguishable phenotypes.

Even though ambulation through an open field or movements back and forth in a shuttle box are the most frequently employed measures, there is a subtle difference when these Behaviours are used to investigate exploration rather than activity. One way to separate *exploratory* activity from general activity is to measure ambulation for brief (usually 5-minute or 10-minute periods) single trials, or to count only those movements occurring when the subject is first placed in the apparatus.

The assumption here is that exploratory Behaviour is the activity that occurs in a novel environment. Another method is to measure different *kinds* of movement (*e.g.*, rearing, as opposed to ambulation), especially when barriers or other objects are placed in the novel environment. Birke provides a good recent discussion of the aforementioned problem, and offers a similar method (using specific object exploration) for improving the measurement of exploratory activity.

The likelihood that these changes in procedure have resulted in measuring something other than "mere" activity is supported by the nature of the genetic models suggested. Although polygenic systems seem to be the rule for activity it is not uncommon to find that exploration is controlled by a single locus.

VAN ABEELEN

Two inbred strains of mice (SRH and SRL) were developed from two lines originally selected for high and low frequency of rearing Behaviour in a novel environment. The novel environment consisted of an open field that sometimes contained an empty food hopper. Van Abeelen created two backcross populations by crossing the high rearing frequency SRH and the low rearing frequency SRL mice, respectively, with mice from an inbred strain (DBA/2; one of the progenitor strains for the original selection study).

When he then tested the intercrosses (offspring of the two backcross generations) in the novel enviomment, the rearing scores formed a clearly bimodal distribution, indicating the likelihood that a single locus controls this phenotype. There was also a bimodal distribution of open field activity scores, although not as striking as that obtained for rearing frequency.This, together with the high correlations between rearing frequency (vertical activity) and open field movement (horizontal activity), led van Abeelen to conclude that the two phenotypes are both under monogenic control, and that: "The genetic factor might be allelic with the gene *Exa* [affecting "exploratory activity level"] discovered by Oliverio, Eleftheriou, and Bailey."

It does not seem farfetched to consider rearing up to investigate even a familiar stimulus (an empty food hopper) in a novel environment as a form of exploration. The argument in favour of a single locus controlling activity, and a necessary causal connection between activity and exploration, is considerably less compelling. First, mice ambulating in a novel enviomment are not necessarily being active for the same reasons as those ambulating through a more familiar environment.Second, if mice of two genotypes differ to the extent to which they will rear up to investigate an object, they would also be apt to differ in the horizontal activity needed to approach these objects initially. (We assume they do not walk on their hind legs.) Thus, the "causal relationship" between exploration (rearing) and activity (open field ambulation) could well be an artifact of the experimental setting. Non-etheless, van Abeelen's studies do provide a distinctive approach to the genetic analysis of one type of exploratory Behaviour.

OLIVERIO, ELEFTHERIOUS AND BAILEY

Using a completely different technique for genetic analysis, and also a different Behavioural testing procedure, Oliverio et al. also reported that what they termed "exploratory activity" is controlled by a single locus. The genetic technique they used involved the use of recombinant inbred (RI) strains and congenic histocompatibility lines of mice to establish a precise genetic model. In the Oliverio mice of the seven CXB (Bailey) RI strains, together with their C_{57}BL/6By (B6) and BALB/cBy (C) progenitors and reciprocal F_1 hybrids were tested in a small shuttle box for 10 minutes. Shortly before testing, each subject received an intraperitoneal injection of saline. The scores on "short term exploratory activity" were measured by crossings from one end of the shuttle box to the other.

The results yielded a strain distribution pattern, confirmed by the testing of a congenic histocompatibility line, showing that short term exploratory activity is controlled by a gene linked to the *H*-26 locus on chromosome 4 (Linkage Group VIII). In this same study, they found that exploratory activity is changed (for several strains, reversed) by the effects of increasing dosages of scopolamine. A single gene controls the changes in exploratory activity due to the administration of scopolamine. Changes in exploratory activity following injection of amphetamine were found to be due to a polygenic system. In a later study, the genetic model for short term exploratory activity (following a saline injection) was replicated, and the locus for its modification by ethanol was found: *Eam,* linked to the *H-16* locus on chromosome.

These studies are important for several reasons. The 1973 study was one of the first to make use of recombinant inbred strains (and the related techniques) to establish a specific genetic model for a *Behavioural* phenotype. Unfortunately, the influence of these same studies on our understanding of exploratory Behaviour has been largely negative, due to the misapplication of the phenotype.

The 1973 paper has been widely cited, and the gene discussed in that paper was named *Exa*, which has led to the conclusion by some that the locus controlling exploratory Behaviour (or exploratory activity) has been determined.

But were Oliverio et al. measuring exploration? It seems unlikely that either exploratory Behaviour or general activity is involved to any great extent in whatever Behaviour is controlled by *Exa*. It hardly seems credible that a small shuttle box, hardly longer than the lengths of three mice, could serve as a novel environment for as long as 10 minutes, or even a fraction of that time. Evidence against considering shuttle box activity to be a measure of exploratory Behaviour on stimulus reactivity.

Satiation to much more complex settings has been shown to occur within that time span in several species. It would also be going beyond the data to term shuttle box crossings general activity, as Oliverio et al. do, in the absence of correlated measures. What Oliverio et al. could well have been measuring instead of exploratory activity might have been an activity response to stress or painful stimulation. Remember that shortly before testing, each subject received an intraperitoneal (IP) injection of saline.

We attempted to replicate this study in our labouratory, using the same strains of mice and the identical apparatus and procedure, but *without an IP injection*. The results we obtained were vastly different, with not only many differences in the rankings of the strains, but also a strong indication of a polygenic effect.

There is other evidence that IP injection can have major effects on Behaviour. For example, blank IP insertions can greatly mitigate the severity of audiogenic seizures in DBA/2 mice. There seems to be sufficient contrary evidence to dispute that *Exa* is directly concerned with either exploratory Behaviour or exploratory activity.

EXPLORATION OF NOVEL STIMULI

Its movement in a particular apparatus—but no mention was made as to why the animal moves about. It could be a spontaneous Behaviour as a result of physiological changes in the animal—one theory (drive reduction) includes this presumption. Perhaps there is something of "interest" to the animal that motivates it to explore its surroundings. Before discussing exploratory Behaviour studies, it is important to understand the various components that are involved.

Stimuli have several properties that can be varied to affect Behaviour: novelty, complexity, and the interaction of these two. The novelty of a stimulus is quite important in the initiation of exploratory Behaviour of that stimulus. The more novel a stimulus, the greater the likelihood that it will be approached and examined by the animal. Given a choice between two stimuli, the subject generally will choose to examine the more novel of the two.

There appears to be a bell shaped curve in the effect novelty has on exploratory Behaviour. There is a degree of intermediate novelty that produces maximum exploration; too much or too little and the animal becomes indifferent to the stimulus or spends less time examining it One characteristic of novelty is that of change; the changing of a familiar stimulus makes it more novel.

This ties in with the concept of satiation or familiarization. If the animal is familiar or satiated with the stimulus, it will normally spend less time exploring it. If, however, some aspect(s) of the stimulus is/are changed, the animal will examine it for a longer period of time. The satiation of the subject to a stimulus or environment poses a problem in the measurement of exploratory Behaviour. As the trial or observation continues, the stimuli or environment involved in the experiment becomes more familiar and less novel until it loses its novelty almost completely.

So the Behaviour recorded at the start of the trial is more likely to be indicative of exploratory Behaviour than is the Behaviour at the end of the trial. Studies by Berlyne, Montgomery, and Thompson and Solomon have shown that exploration declines with time in a trial where the subject is exposed to the same object or environment for some specified period of time. Thompson and Solomon also found that when rats were confined in a limited space (to which they were not habituated) for a period of time, exploration initially increased before it began the ultimate decline.

This could be due to an interaction between fear and exploration and makes studies using such techniques more vague in the measurement of exploratory Behaviour. Using rats as subjects, he projected pairs of pictures of animals side by side on a screen for 10 seconds, for 10 trials. One animal in the pair reappeared constantly on one side, whereas the other side had a different animal projected each time. Berlyne found that the subjects spent more of the 10 seconds fixating on the novel pictures on the varying side than on the recurring picture. Once the organism approaches the stimulus, the complexity of the stimulus becomes a major determinant of attention and exploration by the subject. If two stimuli are both novel, the subject will generally react more to the more complex of the two. A complex stimulus contains more variety than simple stimulus, and thus provides more ways for the subject to group the various components and come up with different responses. Several components seem to comprise the attribute of complexity (all other things being equal): the number of distinguishable elements; an increasing dissimilarity between those elements; and an inverse relationship between complexity and the degree to which several elements are responded to as a unit.

That is, the perceptual organization of the stimuli by the subject has a bearing on its degree of complexity. Clearly, the novelty and the complexity of a stimulus will integrate with each other to give each stimulus a particular characteristic. Both are variables that influence exploratory Behaviour, but the methods for manipulating each are different. The modification of a previous stimulus results in novelty and implies that the organism has an expectancy about the stimulus that is not fulfilled (when it views the modified novel stimulus).A succession of events (stimuli) containing item(s) with short term novelty will probably be more complex during the presentation than a succession that is purely repetitious. In addition, "a pattern with a high degree of synchronous complexity will probably have a high degree of relative novelty." Assessing the involvement of novelty and complexity in experiments is quite difficult, as these two variables generally are confounded.

Dember, Earl, and Paradise offer a nice illustration of both variables. One of the pathways had its walls lined in black and white horizontal stripes, and the other walls of the second pathway were either homogeneously black or white. Sixteen rats were run for 60 minutes per day for 5 consecutive days. They found that on the first day, eight rats preferred the more complex pathway (the one with the horizontal stripes).

On Day 2, four more rats preferred the complex pathway, and the last four rats shifted preferences on the third day. None of the subjects regressed to a preference for the plain pathway. These results indicate that "any change in preference will be from the less to the more complex path.

It can be seen that even though novelty and complexity are operationally different, they can be considered to be similar in their effects. The environment in which an organism responds is as much a complex of stimuli as are the stimuli to which it is presumably responding. If the environment is unfamiliar, the organism may be responding to its novelty rather than the properties of the stimulus.They introduced rats into a T-maze with the goal arms blocked by glass partitions at the choice point. The rats were satiated over varying exposure times and the goal arms were either both black or both white. On the test trial the colour of one of the arms was changed and the glass partition removed, allowing the subject to choose between a black or white alley. The results were as expected; the subjects chose the novel stimulus (the changed arm), indicating a "curiosity motive"as Kivy et al. stated it.

Or it could be that the animals chose the changed arm as a result of being satiated by the other arm. Dember gave support to the novelty of environment hypothesis by reversing the satiation and the test trials of the Kivy et al. experiment. What this points out is that a change in the environment, such as brightness in this case, can result in a different response by the subject.

Dember's experiment predicted and found the same Behaviours experiment, but provided a different interpretation to those findings. This shows that there can be a novel environment or novel stimuli in a familiar environment. Granted, the brightness change in the aforementioned experiment is a stimulus as well as part of the environment.

RESPONSIVENESS TO NOVEL AND COMPLEX STIMULI

Exploratory Behaviour can be viewed from several different perspectives:

approach situations (exploration) and avoidance. Avoidance to novel stimuli generally taps into fear responses and degree of conflict. In these cases, the exploratory Behaviour exhibited by the subject is such that the animal avoids a particular stimulus or environment.

Normally, these studies are confounded, as it is difficult to tell if a subject is not moving because of fear or because of lack of interest. This situation is not as commonly used as approach measurement, but has been used in several studies. Regarding exploratory Behaviour as an approach situation is more common.

In most exploratory Behavioural studies, the animal can either approach a novel stimulus or not, and the resulting Behaviour is taken to be an indication of the degree of novelty and complexity of the stimulus.

Genetic Differences in Exploratory Behaviour

As in activity, genetic differences have been found in exploratory Behaviour. Harrington has done many studies on various measures of exploratory Behaviour in the rat. In runway learning, he has shown that MNRA, INR, TS3, and WAG strains were the fastest. The F344 had very poor runway performance. He also found strain differences in rats in the exploration of different environments, with MNR, ACI, and MNR-a lines having the highest latencies of emergence from a home cage into a novel environment.Harrington brings up the dilemma of confusing a "novel environment" with novel stimuli within a familiar environment and admits that this problem still exists in the literature. McClearn, also using six inbred strains of mice employed four measures of activity: arena, hole in wall, open field, and barrier, to determine the generality of genetic differences.

Again, both strains of C_{57}s were found to be more active than BALB/c mice on all but the hole in wall measures. In 1972, Simmel and Walker demonstrated that exploration of novel stimuli can be differentially enhanced in different strains by the presence of a companion.

We have shown that the investigation of exploratory Behaviour as responsiveness to novel/complex stimuli is certainly not new. Despite this, most Behaviour genetic studies have used activity measures for the investigation of exploratory Behaviour, with the resulting confusion regarding the appropriate phenotype.

REACTIVITY

A technique that permits the direct measurement of the responsiveness of an animal to novel stimuli (*stimulus reactivity*) separately from "general" locomotor activity.

Procedure: It consists of a 20-cm square arena constructed of black Plexiglas. The arena is diagonally bisected by a partition, also of black Plexiglas, with a 1-cm^2 opening in the bottom centre. This allows the subject free access between the two halves of the arena. One-half of the arena (*the novel side*) has murals consisting of black and white diagonal stripes, vertical stripes, and check erboard designs.

Fig. Arena Used for Measurement of Stimulus Reactivity and Activity

The other half (*the plain side*) has no stimuli on the walls. The entire arena is placed atop the sensors of an automated activity metre in such a fashion that identical movements on either half of the arena will yield identical responses by the metre's counter. A 10-minute trial begins when a subject is placed on the plain side of the arena. Two experimenters observe the subject by viewing a mirror angled atop the apparatus.

Operating a counter and a stopwatch, the following measures are taken: Latency (LAT), the time between the start of the trial and the subject's initial entry into the novel side; novel side time (NOV), the number of seconds (of the 600-second trial length) the subject spends on the novel side of the arena; arena crossings (ARX), the number of crossings through the hole in the centre partition between the novel and plain sides; and activity (ACT), the number of movements by the subject, as recorded on the counter of the activity metre.

Under this procedure, it is *possible* to obtain almost any combination of scores on the four measures taken. For example, a subject can be highly active on the plain side without ever venturing into the novel side; or can spend nearly all its time, inactively, on the novel side; or dart quickly back and forth between the two sides.

MULTIVARIATE ANALYSIS OF RESPONSE MEASURES

The first study using this procedure determined the relationship among the several dependent measures. A total of 66 male mice, 50-70 days old, were tested. The mice represented 11 different genotypes. All subjects were tested using the procedure described and, in addition, immediately after this testing, were placed in a shuttle box similar to that used by Oliverio et al. for 10 minutes.The four arena measures (LAT, NOV, ARX, and ACT) and shuttle-box crossings (SHX) were intercorrelated and subjected to a principal components factor analysis. It can be seen that two factors emerged. Two of the arena measures, latency and novel-side time, had high loadings on Factor I (stimulus reactivity) and loadings of nearly zero on Factor II (undifferentiated activity). One arena measure, gross motor activity, had a high loading on Factor II and the lowest loading among the four arena measures on Factor I. Arena crossings had relatively high loadings on both factors. Both measures having high loadings *only* on Factor I (latency and novel-side time) are Behaviours dependent on the subjects' responsiveness to novel stimuli and are thus considered indices of stimulus reactivity. They are also relatively independent of gross motor activity. Arena crossings appears to measure *both* stimulus reactivity and activity. Notice that shuttle-box crossings appears to be independent of stimulus reactivity and to be primarily a measure of motor activity. The factor structure found in the study just described clearly showed that stimulus reactivity (response to novel stimuli) can be distinguished from activity. Next, it was necessary to establish as precise a genetic model as possible, in order to determine the "fit" between the nature of the genetic determinants of the four measures and their loadings on the two factors obtained.

The technique referred to was developed by Bailey and involves the use of recombinant inbred (RI) strains for the determination of strain distribution patterns (SDPs). The Bailey RI strains were derived from a cross of two unrelated but highly inbred progenitor strains, BALB/cBy (C) and C_{57}BL/6By (B6). They are then maintained from the F_2 generation onward by strict inbreeding.

Table. Factor. Loadings of Stimulus Reactivity and Activity Measures

Variable	*Factor I*	*Factor II*
LAT (latency to first novel entry)	-.641	-.011
NOV (novel-side time)	.506	-.122
ARX (crossings from plain to novel side)	.586	.302
ACT (gross locomotor activity)	-.253	.414
SHX (shuttle box crossings)	.206	.616

Inbreeding of the independent RI lines obtained from this cross results in genetically fixing, as full homozygosity is approached, the chance recombination of genes that occurred after the F_1 generation.

Unlike conventional crosses from inbred strains, this technique provides a continuing (and thus, replicable) supply of genotypically identical animals for the investigation of complex phenotypes, enabling one to establish genetic models through use of SDPs.

The seven Bailey RI strains are tested together with the progenitor strains and the reciprocal hybrids. If the progenitor strains differ significantly on the phenotype under investigation, it is possible to compile a strain distribution pattern denoting which of the RI strains resemble BALB/c (C-like), and which resemble C_{57}BL/6 (B_6-like).

It is then frequently possible to match the SDP obtained for the phenotype under investigation with one previously established for distinctive loci, such as histocompatibility. Matching SDPs indicate the possible identity or close linkage of the genes. Because it is possible for two SDPs to match by chance,

$$[p = \frac{17}{2}(=.0078],$$

it is advisable to confirm linkage by testing candidate congenic lines. The congenic (histocompatibility) lines were developed independently from an initial cross of B_6 to C by a regimen of skin graft testing and backcrossing to B_6 for at least 12 generations. This procedure resulted in a battery of congenic B_6 lines, each line being distinguished from B_6 solely by a chromosomal segment that includes a C-strain allele at a distinctive histocompatibility locus.

Each of these congenic lines has been tested against the seven Bailey RI strains to determine which strains carry the B6 allele and which the C allele at a given histocompatibility locus.

Thus, matching an SDP obtained for any given phenotype with an SDP for histocompatibility indicates the likelihood of pleiotropism or linkage of controlling genes at the locus involved. Additional confirmation regarding the specification of the locus in question is derived from testing the congenic line that carries the histocompatibility gene of the matching SDP.If the scores of the congenicline mice are C-like, this would confirm the control of the phenotype under investigation by a gene at this locus, as the mice of this congenic line are identical with B6 at all other loci.The probability of an unlinked gene outside this chromosomal segment after 12 backcross generations is $½^{12}$ or 24×10^{-5}. Whether or not an SDP can be established, it is possible to obtain other sorts of information through the use of RI strains, progenitors, and hybrids.

For example, a continuous distribution of scores on a phenotype among the RI strains, together with a lack of a difference between the progeni tors, indicates that a polygenic system underlies that phenotype.

It is possible to determine dominance by the conventional method of backcrossing hybrids to the progenitors, which differ from each other; sex linkage can be established beginning with the analysis of a difference between the reciprocal hybrids, etc. In addition to the Bailey RIs, several sets of RI lines using various progenitors are currently being developed. The study to determine the genetic analysis of stimulus reactivity tested 132 mice (which, as in the first study, were 60-day-old males), using the same testing procedure as was used for the earlier study, except that the shuttle box was omitted.

There were 12 mice of each genotype of the Bailey RI set. On the two measures that had been found in the first experiment to have high loadings on stimulus reactivity and low loadings on activity (latency and novel-side time), it was possible to establish a strain distribution pattern (SDP), with the same SDP being found for both measures.When confirmed by testing mice from two congenic histocompatibility strains—$B_6.C\text{-}H\text{-}24^c$ and $B_6.C\text{-}H\text{-}1^b$—it was confirmed that both stimulus reactivity measures (LAT and NOV) are controlled by genes at chromosomal segments of $H\text{-}24^c$ and $H\text{-}1^b$ on chromosome 7 (Linkage Group I). The results of these tests on the congenic lines and progenitor strains.

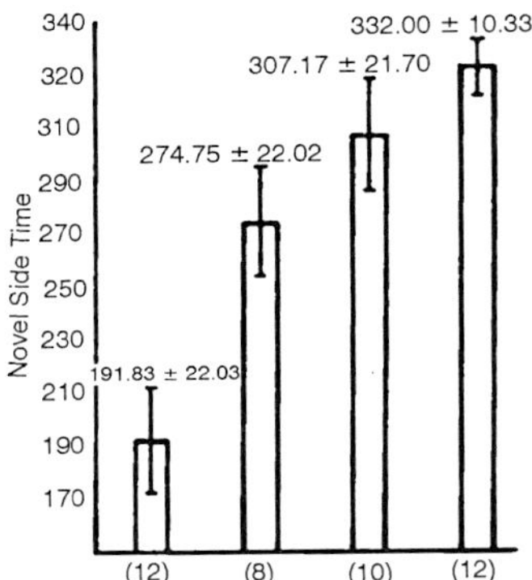

Fig. Comparison of Congenic Histocompatibility Lines with BALB/cBy and C_{57}BL/6By on (a) latency and (b) novel-side time

A polygenic model was found for the two measures with loadings on Factor II (activity and arena crossings), with the progenitor C_{57}BL/6By and BALB/cBy strains not differing from each other on either activity measure.

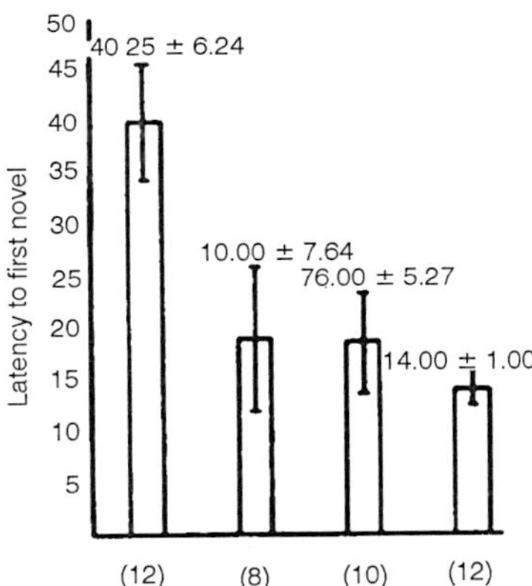

The results of the study show that the genetic model is supportive of the factor structure previously obtained: Exploratory Behaviour measured in terms of reactivity to novel stimuli can be determined separately from general activity and is lines differ from BALB/*c* but not from C_{57}BL/6, confirming the involvement of these loci on stimulus reactivity.

GENERALITY AND STABILITY OF STIMULUS REACTIVITY

The measures obtained in the stimulus reactivity arena appear to be quite reliable over time. The procedures and the general factor structure have been found to be basically similar when species other than mice are tested, with appropriate modifications of the apparatus. Stimulus reactivity and activity measures have been obtained for rats, gerbils, and dogs. In mice, the stimulus reactivity measures have been found to be surprisingly resistant to change due to either subject or environmental variables.

When male and female mice, either 60 or 125 days of age were tested, only genotype differences were obtained on the two stimulus reactivity measures (LAT and NOV), whereas activity differed according to age and sex. In another study C_{57}BL/6By and BALB/cBy mice were reared for several weeks in complex three-tiered cages or were placed in such complex cages one week before testing or were maintained in normal labouratory cages.

IMPORTANCE OF BEHAVIOUR

It has been demonstrated, several decades ago as well as in the current literature, that exploratory Behaviour is a viable phenotype and can be measured as a response to novel and complex stimuli, rather than as a vague form of activity. We suggest that further research on the genetic aspects and ramifications of exploratory Behaviour would be useful, especially in two general areas.

One research area of potential importance is the relationship between exploratory Behaviour and other complex forms of Behaviour. For example, several years ago, Hayes suggested that general intelligence (in humans and in other animals) is an acquired characteristic, resulting from an individual's experience with the environment.

Individual differences between those in similar environments are due to largely inherited differences in "experience producing drives," or the tendency to explore. Another, and perhaps more fundamental topic relates to the adaptive significance of exploratory Behaviour.

Exploration is the form of Behaviour by which an organism is most likely to obtain information about changes in its environment. Humans, rats, and mice, among other species, seem to adapt well to rapidly changing conditions in their environments, and much could be learned about the adaptive processes through further investigation of exploratory Behaviour.

8

Pharmacogenetics

INTRODUCTION

Individual differences in both physiological and Behavioural responses to pharmacologic agents have been known for many years; the sources of these differences are often assumed to be a complex constellation of biological, environmental, experiential and for humans, sociological factors. In one sense, drug responsivity can be viewed as following a continuum; a given dose of a given drug may elicit little or no response, a moderate (often desirable) level of the same response, or an extreme response in various individuals of a patient or subject population.

These variations might be considered quantitative differences in the magnitude of a given response following drug administration. Descriptive statistics, such as ED50 (the dose of a drug that will be effective in producing the desired response in 50per cent of the subject population) and LD50 (the median lethal dose), are sometimes useful indices of the average response to a particular drug. However, the necessity of using such indices of central tendency is evidence for the inherent variability in drug responsiveness among individuals in a given subject population.

A second category of individual differences in response to drugs is evident when a proportion of individuals in a population display a qualitatively unique response to even moderate doses of the drug. This second category has often been called adverse or idiosyncratic drug reactions.

An example of an adverse reaction is the hemolytic anemia produced by primaquine and related drugs in certain individuals with glucose 6 phosphate dehydrogenase deficiency. Such idiosyncratic adverse drug reactions can come about through abnormal ways in which the body acts on drugs or via abnormal states of the body that are revealed only upon drug challenge. Documentation of idiosyncratic drug reactions is prob ably greatest for drugs used in therapeutic settings, examples of which are discussed later, and more complete reviews of which are presented elsewhere. Pharmacogenetics is an emergent discipline that seeks to elucidate the contribution of genetic factors to the types of individual differences in drug responses presented in the preceding paragraphs. The related discipline of Behavioural pharmacogenetics stems from the realization that Behavioural responses to drugs are no different in principle from physiological or biochemical responses, and further that altered biochemical processes often can be reflected in variations in observable Behaviours.

The principles and methods of Behaviour genetic analysis apply equally for evaluating differences among genotypes of mice in an open field, regardless of whether or not the mice have been previously treated with a psychoactive substance. Similarly, the methods used to attempt to describe the relative importance of genetic and Non-genetic factors for individual differences in specific cognitive abilities among humans can also be applied when the phenotype in question is one related to drug responsivity (*e.g.*, differences in responses to acute or chronic administration of alcohol).

That is, the roots of Behavioural pharmacogenetics are based in the older disciplines of psychology, genetics, and pharmacology. In its evolution, the field passed through the interdisciplinary fields of Behaviour genetics, Behavioural pharmacology, and pharmacogenetics.

In describing the nature of the then emerging discipline of psychopharmacogenetics, Fuller and Hansult and Eleftheriou referred to its hybrid nature. Fuller and Hansult went on to ask how far we might extend the metaphor by asking the relevant question: "Is this simply an area of overlap between three established areas or does it have characteristics of its own, even a touch of heterosis or hybrid vigour" Some years later, Horowitz claimed that heterosis had indeed occurred. However, he was referring to heterosis in the context of the rapid development and increasing numbers of researchers with interest in the area.

The field has begun to be realised, as evidenced by the summaries provided by Broadhurst and others. However, it has yet to be established if heterosis, in the sense that Fuller and Hansult intended, has been realised in the field of Behavioural pharmacogenetics. Does the field, in fact, have characteristics that are unique and independent of its progenitor disciplines? In the following discussions of critical issues and future directions, we hope to offer suggestions for directions that might result in this latter type of heterosis.

Until we address those issues, however, we should like to be bold enough to introduce a new metaphor to describe the progress of Behavioural pharmacogenetics. We refer to it, not only as a hybrid science, but also as a fully segregating science. The metaphorical question thus becomes whether or not the elements derived from the dihybrid cross follow Mendel's law of independent assortment. We think that they have, and that this independent assortment is evident in at least two divergent strategies that have emerged within the broader goal of Behavioural pharmacogenetics. This broader goal is the demonstration that indi vidual differences occur in Behavioural responses to drugs, and that genetic factors must be shown to contribute to these individual differences. This primary goal has been realised by Behavioural geneticists and pharmacologists alike. As McCleam has stated, we could probably find strain differences in virtually any trait of interest to Behavioural pharmacology. The two strategies referred to differ as to how to proceed from this initial goal.

One approach is more concerned with elucidating the genetic architecture of the Behavioural response to a given drug. Its advocates suggest the application of such tools of Behaviour genetic analysis as classical genetic breeding designs, diallel crosses, and family studies, to learn more about such genetic parametres as mode of inheritance, heritability, and phenotypic and genetic correlations.

The second approach uses the tools of genetics to produce (*e.g.*, via selection) or to find (*e.g.*, via strain surveys) populations of subjects that differ in their Behavioural responses to drugs, and then to look for underlying physiological or neurochemical mechanisms that might mediate these differences.

It should be noted that the differences between the two strategies is not so much in the initial methods employed, but rather in the interest and subsequent course of further investigation.

The former approach, here termed a biometrical approach, was the thesis for Broadhurst's recent survey of Behavioural pharmacogenetics. In this work, Broadhurst offered an admirable survey of the research to date in this field.He further suggested that careful examinations of the genetic architecture of the traits in question have rarely and only recently been conducted.

Therefore, we concentrate on the contributions made to the field by researchers employing the second approach, that of attempting to define possible physiological and neurochemical substrates of genetic differences in Behavioural responsiveness to psychoactive agents. It should become apparent, however, that the two approaches are not always separable in practice.

HUMAN PHARMACOGENETICS

Ecogenetics is a field of research defined by Omenn and Motulsky as the study of genetically determined differences in susceptibility to environmental agents. Although this broad definition includes such interesting topics as nutritional requirements and carcinogens, is the field of pharma-cogenetics. Both therapeutic and recreational drugs have effects that show considerable individual variation. A thorough review is provided by Omenn and Motulsky. We present a few examples here for the purpose of describing the kinds of pharmacological agents and genetic mechanisms that might be involved in these individual differences.Important therapeutic drugs such as antipyrine, phenylbutazone, and halothane are known to demonstrate marked individual differences in the attained levels in blood and in metabolism rate. These pharmacokinetic factors are known to have a considerable basis in hereditary factors, as demonstrated by twin and family studies. Genetic control of pharmacokinetics of drugs used in psychiatric therapy is well demonstrated by the polygenic nature of inheritance of nortriptyline kinetics. Response to two classes of antidepressants, MAO inhibitors and tricyclics, is thought to reveal two genetically specific types of depression.

Responders to MAO inhibitors generally respond to all MAO inhibitors, but not necessarily to tricyclics and vice versa. The mode of inheritance was not determined in this work, but the mechanism of action is probably not one of dispositional influences, but more related to primary sites of drug action.

Omenn and Motulsky describe a dramatic condition where the widely used muscle relaxant, succinylcholine, can paralyze breathing for several hours. One in 2500 Caucasians shows this response due to an abnormal variant of plasma pseudocholinesterase, the enzyme responsible for inactivation of the drug.This condition is nearly as dramatic as another in which a single autosomal dominant gene produces susceptibility to malignant hyperthermia induced by inhalation anesthetics such as halothane or ether, and also succinylcholine. This rapid rise in temperature often results in death.

Another question that can be addressed is the issue of a hereditary predisposition for abuse of psychoactive drugs. The answer to this question may lead us on a search for "addictable" physiologies or, perhaps, personalities. Clearly, ethanol is a drug with widely varying individual effects. Vesell and Passananti demonstrated high heritability for rate of ethanol elimination in twins ($H = .98$).Yet the fact that alcoholism runs in families and probably has a genetic component is not likely to be explained on the basis of metabolism rates alone. Recent work by Schuckit, Goodwin, and Winokur not only demonstrates a reliable genetic component in alcoholism, but may be interpreted as indicating that environmental influences of sharing a household with an alcoholic are of less importance than sharing common genes. The question of etiology of this as well as other addictions may centre on the aforementioned "addictable" physiologies and/or personalities.

NON-HUMAN PHARMACOGENETICS

The majority of studies using Non-human subjects reviewed herein were designed to examine at least possible differences among groups of genetically defined subjects in Behavioural responses to various psychoactive drugs. However, it should be noted that individual differences are almost always observed, and have often been reported with interest by some investigators using random bred populations. For example, Goldstein and Sheehan found large variations among Swiss—Webster mice in levels of locomotor activity following activating doses of the opiate agonist, levorphanol.

These authors speculated that genetic factors contributed to the observed differences. Perhaps these early observations were the impetus for a selective breeding study conducted in the same labouratory several years later. Using Swiss—Webster mice as the founding population, Judson and Goldstein successfully selected for quantitative differences in changes in locomotor activity induced by an acute dose of this same opiate agonist.

Cappell and LeBlanc investigated the role of morphine pretreatment on later ingestion of a morphine adulterated solution in randomly bred Wistar rats. Individual differences in acceptance of the morphine adulterated solutions were observed in both premedicated and control groups. Those rats in the pretreatment group who were classified as "Non-drinkers" could be induced to drink morphine solutions by subsequent injections of morphine and finally reached a level of morphine ingestion equal to the amount consumed by "drinkers" in the control group.

Other studies using different populations of rats have shown clear genetic differences in acceptance of morphine adulterated solutions. Although the differences in acceptance of morphine solutions observed by Cappell and LeBlanc were not established to be of genetic origin, their results are compatible with a possible genotype by treatment interaction.

Individual differences in drug responsivity have been reported also in similar populations of randomly bred mice. Ho, Loh, and Way reported significant differences in morphine induced analgesia, lethality, tolerance, and physical dependence in Swiss—Webster and ICR mice.

On several measures; there were significant differences among different populations of Swiss—Webster (or ICR) mice when these populations were purchased from different sources. Thus, although individual differences in drug responsiveness among animals of randomly bred populations are often sources of hypotheses that can be tested via genetic or pharmacologic methods, these differences may also contribute to uncertainty when attempts are made to replicate results from different labouratories.

PERSPECTIVES AND ORGANIZATION

In Non-human subjects, genetic factors have been shown to influence responsiveness to virtually every class of psychoactive agents, including (but not limited to) stimulants, mild tranquilizers, barbiturates, alcohol, anesthetics, nicotine, opiates, and neuroleptics. Much of this work has been reviewed or collected previously by Broadhurst and Eleftheriou. Similarly, reviews of humanpsy chopharma-cogenetics have also been compiled.

In keeping with the title of this volume, we have concentrated on the principles and applications of the methods of Behaviour genetic analysis with respect to their relevance to the study of individual differences in Behavioural responses to drugs. Specifically, we have concentrated on research using alcohol and opiates. It is hoped, however, that the principles, applications, and implications that we discuss will extend to the study of the Behavioural effects of any drug of choice by a given researcher.

Our decision to limit much of our discussion to the pharmacogenetics of alcohol and opiate responsiveness goes beyond our own research interest in these two classes of drugs. First, these two classes of drugs are important from a sociomedical perspective; both alcohol and the opiates (primarily, heroin) remain major drugs of abuse. Secondly, alcohol is probably the most widely studied drug in Behavioural pharmacogenetics, and research in the pharmacogenetics of opiates is steadily growing.

Finally, there has been an increased interest in recent years in possible commonalities and interactions between alcohol and opiates. First, we present the major techniques of genetic analysis of drug responsiveness, for both humans and Non-humans. Secondly, we define phenomena of interest from a pharmacologic perspective and give examples that illustrate each from a pharmacogenetic perspective. Evaluation and interpretation are presented and discussed. Finally, we offer some personal impressions of several directions in which the field of Behavioural pharmacogenetics might be fruitfully extended by future research.

TECHNIQUES

HUMAN STUDIES

The genetic methodologies in human pharmacology are the typical approaches in human Behaviour genetics. They may range from pedigree analysis to adoption studies and segregation analysis. We emphasize one approach that has been useful in the study of alcoholism. Detailed treatment of other methodologies may be found in standard sources. Recent work by Schuckit has emphasized a search for biological markers that might be identifiable in individuals genetically at risk for alcoholism.

Towards this end, individuals from families with a history of alcoholism respond differently on subjective reports of degree of intoxication in that they report feeling less intoxicated than controls. Physiological response to alcohol shows similar differentiation. More interesting, though, is the report that men with familial history of alcoholism exhibit higher levels of circulating acetaldehyde after ethanol consumption than controls. It is possible that some of the addictive properties of alcohol are actually attributable to its metabolite acetaldehyde.If so, the finding of higher blood levels may be more than just a marker. It may be a prime factor in the etiology of alcoholism. At any rate, the need for such markers is obvious in order that individuals at risk for alcoholism can be closely studied. In this way those environmental factors that precipitate alcoholism may be uncovered with the use of a genetic tool.

NON-HUMAN STUDIES

The major source of Non-human subjects in Behavioural pharmacogenetic research, as is the case in Behavioural genetics in general, has been inbred strains of mice and, to a lesser extent, rats.

The process of inbreeding leads to individuals who are more closely related than are members of a population at large. In the development of inbred strains, inbreeding is secured by sequential and continuous brother—sister mating. After 20 generations of brother—sister mating, virtually all loci will be fixed for alleles in the homozygous state. Thus, all members within an inbred strain are virtually uniformly homozygous and for practical purposes, are genetically identical to each other.

However, the specific alleles that are fixed through the process of inbreeding are randomly determined (with the exception of those combinations that are eliminated by selective pressures). Therefore, although individual members within an inbred strain are (within limits) genetically identical to other members of the same strain, in all likelihood they will differ from members of a different inbred strain in the alleles fixed at many loci.

In the simplest model of quantitative genetic theory, the observable phenotype of an individual is the additive effects of its genotype and its environment. Given that two different inbred strains are reared and tested in the same (within practical limits) environment, then differences between strains on some observable trait (*e.g.*, alcohol preference) must be determined in large part by the different alleles fixed within each. Thus, differences between two inbred strains that are significantly greater than differences within each strain (which may be due to such factors as a lack of homozygosity, microenvironmental differences, and errors of measurement) are usually construed to represent a genetic substrate for the observed phenotypic difference.

Differences between inbred strains is a useful initial demonstration for both strategies of Behavioural pharmacogenetic research. Strain comparisons have yielded significant differences in, and hence a genetic contribution to, such alcohol related responses as in preference for alcohol in a choice situation acceptance of alcohol using a forced ingestion procedure in the hepatic metabolism of ethanol and in neural sensitivity to ethanol as well as in numerous other Behavioural and biochemical phenotypes.

Similarly, genetic factors have been shown to contribute to differences among mice in preference for morphine adulterated drinking solutions in acceptance of morphine solutions using a forced ingestion procedure in readdiction susceptibility in opiate induced analgesia and hyperactivity in neurochemical responses to opiates and in the development of physical and Behavioural dependence on opiates.For the investigator interested in further elucidating the genetic architecture of the alcohol or opiate related phenotype, these initial strain comparisons are usually extended via the tools of Behaviour genetic analysis, such as classical Mendelian breeding designs and analysis, diallele crosses and subsequent analysis, or the use of recombinant inbred strains to gain further information regarding the genetic parametres of the trait in question.

Although these analytical designs and tools have been applied infrequently in traits of pharmacogenetic interest, the reader is referred to Broadhurst for a survey of these attempts and their results.

With respect to the second general strategy of Behavioural pharmacogenetics, the demonstration of strain differences often leads to hypotheses regarding the relationship between a trait and an underlying physiological or biochemical mechanism, or a relationship between two Behavioural traits. For example, C_{57}BL and DBA mice are known to differ in their preference for 10 per cent ethanol solutions when it is offered versus tap water in a two bottle choice situation.

These two strains also differ in the activity of hepatic enzymes of ethanol metabolism. The question now becomes whether or not these traits are physiologically or genetically related. Historically, attempts to answer such questions have relied principally on three different approaches.

The first approach might be considered one of pharmacologic intervention. For example, if the higher preference for ethanol in C_{57}BL mice is casually related to increased ability to metabolize the drug, then interfering with ethanol metabolism should reduce the alcohol preference of these mice. Schlesinger, Kakihana, and Bennett demonstrated that pretreatment with disulfuram (an aldehyde dehydrogenase inhibitor) significantly lowered the alcohol preference of C_{57}BL mice.

However, the magnitude of the decrease was slight. Similarly Ho, Tsai, and Kissin reported that C_{57}BL mice significantly incorporate higher levels of acetylcholine and greater uptake of radiolabeled choline than did DBA mice. Pretreatment with a choline acetyltransferase inhibitor, which would interfere with acetylcholine synthesis, significantly but transiently reduced the voluntary alcohol ingestion of C_{57}BL mice. Although these two pharmacological intervention studies may contribute to our understanding of why C_{57}BLs prefer ethanol, they may tell us little of why DBA mice avoid it.

A second direction that may be taken to elucidate further a difference in two traits in two different inbred strains is to expand the number of genotypes tested. For example, Castellano and Oliverio have reported that C_{57}BL mice respond to morphine injections with increases in locomotor activity, whereas DBA mice do not.

Horowitz et al. have further shown that C_{57}BL mice prefer morphine—saccharin solutions over water, whereas DBA mice in the same situation drink almost exclusively water. Subsequently, Whitney et al. have reported a positive correlation of about. 92 between preference for morphine—saccharin solutions and morphine induced hyperactivity in six different inbred strains. This strong positive correlation was not due to differences in baseline preference for saccharin, nor to differences in level of activity following saline injections. Similarly, we have recently derived a positive relationship between ethanol preference and preference for morphine—saccharin solutions.

With respect to morphine responsivity, Brase tested four inbred strains and two randomly bred lines for hyperactivity following acute morphine administration and the degree of jumping observed following withdrawal from chronic morphine exposure. The correlation between morphine induced hyperactivity and jumping elicited by either abrupt or naloxone precipitated withdrawal was. 93-.96.

Collectively, these correlation strain surveys suggest a possible genetic relationship between the two traits of interest. However, a more stringent test would be to see if whether or not the two traits of interest are covariates in individual mice of a heterogeneous genetic population. However, many of the techniques of pharmacogenetics are invasive, if not destructive.

For example, if one were interested in the possible genetic relationship between morphine induced analgesia and the density of central opiate receptors, the use of heterogeneous mice would require measuring both traits in an individual animal.

Thus, the researcher is forced to accept the possible confounding effect of order of testing. Counterbalancing in this case is not practical, as it is as yet impossible to assay for central opiate receptors without first removing the subject's brain. Ideally, then, a researcher would like the best of both worlds: the genetic uniformity of inbred strains and the variability of a segregating population.

One population of inbred mice that comes close to meeting these criteria consists of recombinant inbred (RI) lines. A fuller description of these mice is presented elsewhere. Basically, two inbred strains are crossed to produce an F_1 generation, which in turn are mated *inter se* to produce the F_2 generation. Beginning with the F_2 generation, a number of brother—sister mating pairs are begun, and brother—sister mating is continued for at least 20 generations. Collectively, the surviving lines represent unique recombinations of the alleles at which the parental types were different. However, within each RI line, all members are virtually genetically identical and all alleles are fixed once again in the homozygous state.

The use of recombinant inbred lines in pharmacogenetic research has recently been reviewed by Broadhurst and by Crabbe and Belknap. The hypothetical example of the relationship between morphine induced analgesia and density of opiate receptors was actually the subject of an interesting application of recombinant inbred lines by Baran, Shuster, Eleftheriou, and

Bailey. These researchers measured morphine induced analgesia and the density of opiate receptors in the brain for 11 isogenic populations of mice: C_{57}BL and BALB (the progenitor strains), the two reciprocal F_1 generations, and seven RI strains.

Their study amounted to a strain survey for each trait in question, as well as a strain ranking correlation between the two traits. Genetic differences were apparent for both morphine induced analgesia, as measured by the tail flick method, and for density of opiate receptors, as measured by stereospecific binding of radiolabeled naloxone (an opiate antagonist).

Finally, although there was a positive strain ranking correlation between binding and analgesia, the differences in opiate receptor density were not great enough to explain a significant proportion of the variation in morphine induced analgesia.

Clearly, properties of opiate receptors are one possible source of genetic differences in the ability of morphine to produce analgesia. However, other factors, including possible differences in neurotransmitter systems modulated by these receptors, are also possible contributors to the phenotypic differences in morphine-induced analgd strains.

SELECTION STUDIES

With inbred strains, the combinations of alleles that are fixed by the inbreeding procedure are random. Thus, the combinations of phenotypic expressions of these genes are also random. By selective breeding procedures, a researcher can select a phenotype of interest and selectively breed for quantitative differences in that phenotype. In research, unlike agricultural applications, it is usually desirable to select for both high and low levels of the phenotype. Different types of selective breeding techniques are available, depending in part on the demands dictated by the desired phenotype. The basic scheme is to start with a heterogeneous founding population and to breed individuals with higher levels to others with higher levels; conversely, individuals with low phenotypic levels are bred with other low individuals. The response to selection is based, in part, on the variability in the founding population, the heritability of the trait in question, the system of selective breeding, and the selective pressure exerted by the experimenter.

Given heritable variance, the appropriate selective pressure, and enough generations of selective breeding, the procedure will produce two lines of subjects that differ markedly in the phenotype undergoing selection.

Historically, a number of selective breeding programmes have been developed for various phenotypes of interest in Behavioural genetics, using both mice and rats as subjects. Many programmes have selected for phenotypes initially not related to pharmacologic interests, but later evaluated for differences in response to various psychoactive agents. These studies are reviewed by

Broadhurst. In addition, several selective breeding studies have been conducted in which the selection criterion was a response to a psychoactive agent; the majority of these studies used some response to alcohol as the phenotype in question.

These studies, including the Colourado sleep lines the most and least affected rat lines the alcohol Non-tolerant and tolerant lines the Finnish Alko Alcohol and Non-alcohol lines and the Indiana alcohol preferring and Non-preferring rats have recently been reviewed, by workers involved in each, in a research monograph by NIAAA.

A successful selective breeding design serves at least two valuable functions. First, the demonstration that selective breeding can produce systematic quantitative changes in a drug related phenotype gives definitive proof of the contribution of genetic factors to individual differences in that particular response. More importantly, such studies often generate hypotheses regarding the selected trait and other traits thought to be related to it.

Ideally, the selection procedure should effect changes in the frequency of only those genes that are related to the response undergoing selection. Therefore, differences among selected lines in some response other than that being selected may be indicative of a physiological or genetic relationship between the two traits in question. For example, rats that have been selectively bred for susceptibility or resistance to readdiction liability to morphine also show predictable differences in their susceptibility to alcohol readdiction. These data are suggestive of a genetically based commonality in the processes underlying addiction to alcohol and opiates.

It is our contention that such demonstrations are valuable in suggesting hypotheses of commonality, but these must be tested by more rigorous experiments. The selection index was duration of the loss of righting reflex ("sleep time") following 3.4 g/kg ethanol, although this dose was increased in subsequent generations of selection. By the fourteenth generation, there was virtually no overlap in the sleep time distribution of mice forming the long sleep and short sleep lines. Subsequent research has shown that these lines of mice differ on a number of other response measures to both alcohol and other agents.

These responses may or may not be genetically correlated to the selection index, but they furnish useful hypotheses regarding the mechanisms of action of Behavioural and physiological responses to alcohol. First, traits for which an *a priori* reason exists to suggest a relationship with the selected trait, may play no realizable role in the product of such selection.

An inverse relationship between ethanol induced sleep time and aldehyde dehydrogenase activity following alcohol challenge in the heterogeneous population from which LS and SS mice were eventually derived. This relationship would lead to the intuitive hypothesis that part of the differential response to selection in LS and SS mice might be due to differences in rates of ethanol metabolism. Yet Heston, Erwin, Anderson, and Robbins reported similar rates of ethanol metabolism in LS and SS mice. A second phenomenon is concerned with the demonstration of concomitant responses in selected lines. LS mice have a longer duration of the loss of righting reflex following hypnotic doses of ethanol than do SS mice. On the other hand, Goldstein and Kakihana demonstrated that SS mice showed a greater severity of seizures, relative to LS mice, following withdrawal from 3 days of chronic exposure to ethanol vapours, even though blood alcohol levels were maintained at approximately the same level in both lines.

These results might be interpretable as demonstrating a relationship between acute tolerance to, and acquired dependence on ethanol. However, in the same paper, these authors further demonstrated the independence of ethanol induced narcosis and severity of seizures induced by withdrawal from chronic alcohol exposure by showing that the two traits were not covariates among individual animals of the original heterogeneous founding population from which LS and SS mice were derived.Thus, the concomitant responses observed in the first part of their study may have been due to a stochastic association of genes subserving each response.

The final phenomenon regarding research subsequent to the development of selected lines is that, even if hypotheses derived from them are tested and and not confirmed, further hypotheses might evolve from them. Ethanol induced sleep time and severity of seizures following withdrawal from chronic ethanol exposure may not be genetically correlated characters; at least no evidence to support this hypothesis can be derieved from the results of the latter study by Goldstein and Kakihana.

However, LS and SS mice still may represent different genetic populations for the withdrawal phenotype. Horowitz and Allan investigated withdrawal from chronic mor phine exposure in LS and SS mice. Briefly, mice of both lines were implanted with sustained release morphine pellets and were withdrawn 3 days later via naloxone challenge. On two classic measures of opiate withdrawal (wet dog shakes and jumping), SS mice were more sensitive to withdrawal from chronic morphine administration.

These results were consistent with those reported for severity of seizures following chronic withdrawal from alcohol exposure, where SS mice again showed the more severe score on one index of alcohol withdrawal.

Two cautions must be taken regarding the interpretation not only of this latter study, but of all tests of concomitant responses in selected lines. First, the results may be specific to the methods and Behavioural assays employed. SS and LS mice may or may not differ with respect to their general level of physical dependence on alcohol or morphine following these drug regimens.Secondly, as Satinder and others have pointed out, the true test of a genetic relationship between two traits is to see whether or not the traits are covariates in individual animals of a population of heterogeneous genetic stock. Thus, the relationship suggested by the results of the studies by Goldstein and Kakihana and Horowitz and Allan remain to be evaluated.

HETEROGENEOUS STOCKS

Genetically heterogeneous stocks are found in several different forms, each of which has its peculiar advantages and disadvantages. The common characteristic of these stocks is a lack of genetic uniformity from animal to animal.

Within that definition, however, is a wide range of genetic diversity. The presence of this diversity provides several tools for the pharmacologist interested in the control of but not specific investigation of genetic variables. For the geneticist, these stocks have a wide variety of uses. First, we distinguish several types of stocks available.

"Random Bred" Stocks

Many commercial suppliers of labouratory animals, particularly rodents, characterize their populations as random bred. The extent of genetic variability in most of these stocks is not known. In house breeding colonies maintained by investigators are likely to be rather inbred owing to founder effects.

This fact and general lack of information on ancestry of these types of stocks make them less suitable for genetic work than the other types of stocks described here. However, Rice and O'Brien have characterized the often used Swiss mouse.This work suggests more genetic variability (enzyme polymorphisms) in this stock than is suggested by the foregoing. The usual lack of controlled breeding schemes and the many different populations of Swiss type mice call into question the generality of this finding.

Inbred Strain Crosses

The production of an F_2 generation from two inbred strains is relatively inexpensive monetarily, as well as in time, and provides a heterogeneous stock with an amount of genetic variability determined by the number of allelic differences in the two parental stocks. Maintenance of such a stock by production of further segregating generations could be accomplished by a random mating scheme with some care taken to avoid inbreeding.

Such a stock with controlled breeding would be suitable for a variety of genetic analyses. The degree of usefulness is limited by the number and character of differences between the two parental stocks. More than two alleles at a locus would not be possible. Choice of progenitor stocks should attempt to maximize genetic diversity for the phenotype(s) of interest or in general characteristics if genes underlying such phenotypes are not identified. The latter approach would simply suggest using maximally unrelated strains.

The Eight or Four Way Cross

The limitations suggested for the F_2 type stocks may be overcome by complex crosses involving more than two parental stocks. Such crosses have several times served as the foundation stock for selective breeding experiments. Their usefulness can, however, extend beyond this role. McClearn, among others, has suggested the use of the eight way cross. Beginning with eight inbred strains of diverse origins it is possible to construct a stock with an extremely high degree of genetic diversity, expressed in a variety of phenotypes and specifically identifiable in numerous enzyme polymorphisms.

It is conceivable that within such a stock, allelic combinations might arise that would not be possible in natural populations. This could occur if some of the progenitor strains were derived from completely separate populations. For example, recently inbred strains from *Mus musculus castaneus* and *M.m.molossinus* both show allelic variants at some loci that do not exist in any of the more commonly used strains.

Such uniqueness might very well provide useful tools for the analysis of drug action. This notion is analogous to the argument that single gene mutants may provide information on "normal" Non-mutant physiology. In this case however, the degree of phenotypic aberration may provide either qualitative or quantitative uniqueness of response.

Experimental Advantages

For the pharmacogeneticist, the major advantage of a genetically heterogeneous stock may be the appropriateness for investigations of genetic covariance or correlations. These correlations may involve several Behaviours affected by a single pharmacological agent or single phenotype effects of several agents. Although the number of subjects required for such analyses may be large, and vary in a negative fashion with character heritabilities benefits of such work may be large. In the present context, heterogeneous stocks provide a clear advantage over the use of inbred strains in such analyses. The fact of complete homozygosity in such inbred strains produces a physiologically aberrant situation. It is conceivable that pharmacological effects might depend not only on allelic genotype but also on the fact of homozygosity. More importantly, the random fixation of allelic pairs as a result of the inbreeding process may have produced unique polyallelic combinations.

To the extent that these genotypic profiles might produce correlated Behavioural/pharmacological effects, potentially large genetic correlations could result. These relationships might however be transient effects that could disappear or diminish in a random breeding population where they would occur with a smaller frequency, as those genotypic combinations would be rare. The presence, then, of genetic correlations in heterogeneous stocks indicates that they are probably more general phenomena.

Another experimental approach that finds powerful application has been termed the genotypic method in Behaviour genetics. The identification of an allelic variant can allow assessment of the pleiotropic effects of such on Behavioural response to drugs. This approach is most often used with coisogenic stocks, but could be accomplished in heterogeneous stocks as well. For example one might identify electrophoretic variants of mouse liver alcohol dehydrogenase, which are known to be under Mendelian control and proceed to assess alcohol responsiveness in mice of differing genotype.

If the phenotype examined were known to be influenced by dispositional variables, then one might expect an influence of ADH. Such pleiotropic effects could be examined for a wide variety of identified allelic variants (*i.e.*, coat colours or enzymes). The advantage of the heterogeneous stock in this case is the ability to examine effects of one genetic variable (the marker phenotype), while maintaining a more or less random genetic background on which it is found.Again, observed effects might be more general than those seen with coisogenic stocks. The utility of heterogeneous stocks for Non-genetic research also deserves comment. Although the advantages of using inbred strains derives from their reproducibility and isogenicity, these variables are also potentially disadvantageous for the researcher not interested in genetic mechanisms. Aside from the generality issue raised by pharmacological work with one or few strains, the fact of genetic homogeneity at all loci may be less than desirable. Simply stated, a completely homozygous mouse is clearly an aberrant type of animal. Inbreeding depression effects may very well extend to pharmacological responsiveness, particularly if the Behavioural phenotype is related to fitness.

The heterogeneous stock provides a physiologically more normal animal. In stocks produced by careful crossing of inbred strains (*i.e.*, 2–, 4–, or 8–way crosses), the genetic ancestry of any particular subject can be as well known as that of any inbred animal. Contrary to popular belief, the variability associated with measurement of a phenotype is not necessarily larger and can be smaller in heterogeneous stocks than in inbreds. The good general health and large reproductive capacity of heterogeneous stocks is also a consideration.

An interesting use of the Colourado HS mice was reported by Reed. In a population of 1055 mice, heart rate was measured before and after administration of ethanol. What occurred was a "normalizing" response. Mice that had deviant baselines scores tended Towards the mean on the postalcohol test. Thus a substantial negative correlation between Pre-ethanol and postethanol scores existed.This normalizing response was also shown to exist in humans. In another study with HS mice it was suggested that effects of ethanol on body temperature, open-field activity and heart rate were not genetically correlated.

MUTANTS AND COISOGENIC STOCKS

The genotypic approach in the preceding as received relatively little attention from the experimental pharmacogeneticist. This is not for a lack of genetic material to work with. Numerous allelic variants of enzymes have been identified in a variety of species. The mouse, as a mammal, is probably the species of choice here as well.

A variety of enzymes could be examined for allelic variant effects on pharmacologic responsiveness. In addition, the presence of numerous neurological mutants provides an obvious subject matter for neuropharmacological if not Behavioural investigation.

Two examples of mutants are pertinent. First, the single-gene neurological mutant, Jimpy, is at least 10 times less sensitive to the analgesic effects of morphine. These mice also have fewer opiate receptors. Whether these effects are direct or executed during development is unknown. A second example is not an identified mutant, but a plausible one. Poley reported that an Alberta subline of the C_{57}BL/6 mouse would not consume ethanol in a choice situation the way that other closely related substrains would. Thus, although both C_{57}BL/6J and C_{57}BL/10J mice consumed considerable ethanol, the C_{57}BL/ 6UAE mice did not. Whitney and Horowitz replicated that work, but also showed that the Alberta subline would consume as much sweetened morphine as would the other sublines. The difference in ethanol consumption is certainly suggestive of a single-gene difference.

PHARMACOLOGICAL PHENOMENA

INITIAL SENSITIVITY

Naive sensitivity to alcohol may be assessed in a number of ways depending on the Behaviour or process of interest. Loss of the righting reflex in the mouse has been a useful phenotype for genetic study. Although some variation in procedure occurs from lab to lab, the basic method is that of Kakihana, Brown, McClearn, and Tabershaw.

After injection, the mouse may or may not be assessed for the length of time needed to fall from a suspended grid (fall time). The primary measurement is the time it takes the subject to right itself spontaneously three times in 30 seconds. This measure gives a general estimate of hypnotic potency of a depressant drug. When combined with assessment of blood levels of the drug at time of waking, an estimate of neurosensitivity is obtained. This, of course, was the criterion phenotype for the selective breeding programme for the LS and SS lines. Extensive study of strain variation in this response to alcohol also exists.

Measures of intoxication are numerous. The tilting plane test, the grid test, and the Rota-Rod have all been utilized in pharmacogenetic work. The grid test involves suspension of a hardware cloth grid over a metal plate. Placement of a paw through the grid onto the plate results in a circuit activation and can be registered as a count.The more intoxicated a mouse, the greater number of "step-throughs." When combined with simultaneous assessment of locomotor activity, a relatively sensitive measure of intoxication can be obtained. That approach determined LS mice to be relatively more intoxicated by subhypnotic doses of ethanol than were SS mice. Dose-dependent effects may be obtained but, at times, only if a correction for general activity levels is made.

A third measure of initial sensitivity examines the purported biphasic shape of the ethanol dose-response curve. Locomotor activity is generally stimulated by low (1-2 g/kg) doses of ethanol. Higher doses of ethanol produce depression. The stimulant properties of low doses are of interest in that they seem to present a paradox: stimulant actions by a depressant drug.

The genetic studies of this Behaviour at times indicate quantitative genotypic differences and in other cases qualitative differences. For example, both LS and SS mice are activated by 1.0 g/ kg, but higher doses depress LS mice and continue to activate SS mice. Alternatively, BALB/c mice show activation to several low doses and C_{57}BL/6 mice apparently cannot be stimulated at all. These examples were chosen to indicate that if an understanding of underlying mechanism is of interest, then it is probably best to study simple Behaviours. Study of complex Behaviours, although important, is difficult from a genetic viewpoint because more genes, and thus more physiological systems, are probably involved.

TOXICITY

Both alcohol and opiates should be seen as toxic agents. Cirrhotic livers and heroin overdoses are *prima facie* evidence. As such, standard toxicological techniques can be applied. A study by Whitney and Whitney showed mortality rate among 10 genotypes to be negatively related to their free choice preference for ethanol. Also, the LD50 values for LS and SS mice were equivalent in mice from the 14th generation.

Another type of toxicity is studied in the conditioned taste aversion experiment where ethanol can serve as a UCS or poison to produce flavour aversions. Towards this end, Horowitz and Whitney showed DBA/2 mice to form much more severe conditioned aversions, when ethanol was the UCS and saccharin flavoured water the CS, than did C_{57}BL/6 mice. Of similar interest is a recent report that BALB/c mice do not form conditioned taste aversions when doses of ethanol up to 6.0 g/kg are used as the UCS.

Chronic Drug Tolerance

This refers to the waning of responsiveness with repeated administration of a drug. It can be expressed in two ways. First, tolerance is evident if the magnitude of the response decreases with repeated administration of the same dose of the drug. Alternatively, tolerance can be demonstrated if, with repeated exposure, greater doses of the drug are required to maintain a given magnitude of the response being evaluated.Tolerance observed at the Behavioural level may be a result of alteration of drug availability (*e.g.*, metabolic tolerance), adjustments made by the organism to compensate for the effects of the drug (*e.g.*, Behavioural tolerance), or changes that occur at the site of action due to continued presence of the drug (*e.g.*, pharmacodynamic tolerance). Genetic factors at any of these levels could contribute to individual differences in the degree of tolerance to either alcohol or the opiates in both humans and Non-humans. Pharmacogenetic aspects of drug tolerance and dependence, including both alcohol and the opiates, have been reviewed extensively in an excellent recent summary article by Crabbe and Belknap. Thus, one example is presented for genetic differences in the development of tolerance to alcohol and to the opiates. Moore and Kakihana examined the development of tolerance to the hypothermic effects of ethanol in three inbred strains of mice. Animals were injected with 3 g/kg ethanol daily for a period of 8 days. Rectal body temperatures were recorded on the first, third, and last day of drug administration. As a general summary, tolerance developed to the hypothermic effects of alcohol in mice of the BALB and C_{57}BL strains, but not in DBA mice.

In fact, DBA mice may have become more sensitive to the hypothermic effects of ethanol with repeated administration. Although interpretation of the degree of tolerance is complicated by the magnitude of hypothermia induced by the initial exposure to ethanol in mice of these three genotypes, it is clear that genetic factors affected the rate and direction of changes in body temperature resulting from repeated ethanol administration.

Although differences in alterations in alcohol elimination rates were apparent in this study, these metabolic differences were not great enough or consistent with the genetic differences in the pattern of tolerance to the hypothermic effects.Tolerance to the analgesic effects of morphine was evaluated by Oliverio, Castellano, and Eleftheriou. Both initial sensitivity and tolerance were evaluated using the hot-plate technique in the Bailey BALB, C_{57}BL, F, and recombinant inbred lines. Mice were initially tested following three doses of morphine and then made tolerant by four additional administrations of morphine. Tolerance evaluation followed the fifth morphine injection. In general, genetic differences were apparent in both initial sensitivity and acquired tolerance to the analgesic effects of morphine as measured by response to thermal stimulation. However, for both the initial sensitivity and the tolerance test, analgesia was evaluated at several postinjection times. Thus, interpretation of the results is complicated by dose by postinjection trial, which clearly could be influenced by the repeated-measures design.

Dependence

Dependence is a consequence of chronic drug administration in which the drug is needed in order for the organism to maintain normal functioning. Both alcohol and opiates can lead to dependence following repeated administration. As Crabbe and Belknap have pointed out, dependence is inferred by the consequences of terminating drug administration (*i.e.*, withdrawal or the abstinence syndrome).Genetic differences in the degree of dependence on alcohol and morphine have been alluded. Goldstein and Kakihana reported greater seizure severity following withdrawal from chronic alcohol exposure in SS than LS mice, whereas Horowitz and Allan found analogous results in mice of these lines in both jumping and wet dog shakes induced by withdrawal from chronic morphine administration. Similarly, the demonstration by Brase et al. of genotypic differences in severity of opiate withdrawal of the use of inbred strains.

Self-Administration

This is another phenomenon associated with chronic drug administration. It can be demonstrated in a number of different paradigms, but in each case, the organism determines the amount of drug to be administered. Due to the number of subjects required in pharmacogenetic studies using Non-human subjects, the most common techniques involve some variation of oral ingestion.

When the procedure employs an alternative to consuming a drug-adulterated solution or food source, drug preference or drug seeking can be examined. Genetic differences in preference for alcohol and morphine-adulterated solutions among geneticaly defined subjects. A cautionary note should be made regarding the importance of task specificity when evaluations are made regarding the genetic contributions to differences in tolerance, dependence, and self-administration.

These cautions will be expanded in subsequent. First, degree of tolerance for a psychoactive drug is often dependent on the choice of response used as a criterion of strength of effect. Secondly, inasmuch as dependence is inferred from the consequences of drug termination, genetic differences may reflect differences in the response being measured rather than differences in the absolute degree of dependence.

Finally, oral ingestion regimens of self-administration may be confounded by factors unrelated to the neurochemical effects of the drug, such as preabsorptive orosensory cues. Although all these task-related factors may be of interest from a Behavioural pharmacogenetic perspective, such factors must be considered in making statements regarding genetic differences in the more general phenomena under investigation.

LOCUS OF EFFECT

Absorption and Distribution

When psychoactive agents such as alcohol and opiates are administered systemically and when genetic differences are observed in response to such administration, it is tempting to attribute the observed differences to genetically determined differences at the site of action, usually in the central nervous system. However, such conclusions are predicated on the assumption that equal amounts of the drug reach the site of action.

The assumption may not be warranted, if genetic factors contribute to different amounts of the drug reaching the central nervous system via such mechanisms as differential absorption, metabolism, and permeability of the blood—brain interface.

Many careful attempts to evaluate such possible differences in drug distribution have been conducted in pharmacogenetic research involving alcohol and the opiates. For example, Brase et al. measured morphine induced hyperactivity and analgesia (using an abdominal constriction technique) in four inbred and two randomly bred lines of mice.

These researchers also measured brain morphine levels at the time of each test. Strain differences were apparent for all measures, including brain morphine levels. However, brain morphine levels did not correlate significantly with either analgesia or locomotor activity, indicating that the observed differences in these latter two phenotypes could not be ex plained by differences in distribution of the drug. Similarly, Moore and Kakihana demonstrated differences in blood alcohol levels following ethanol administration in three inbred strains of mice, but further showed that these differences could not totally account for differences among these strains in ethanol induced hypothermia.

Finally, Goldstein and Kakihana equated blood alcohol levels in LS and SS mice, but still found differences in the severity of seizures following withdrawal from chronic alcohol exposures.In a slightly different sense, issues of drug absorption may be of primary interest in the investigation of Behavioural phenotypes in pharmacogenetic research.

For example, the substrate for the difference in alcohol preference between C_{57}BL and DBA mice has been the subject of considerable research. Recently, Belknap and his associates have suggested that different mechanisms may underlie the avoidance of alcohol by DBA mice, and the preference that has been consistently demonstrated by most sublines of C_{57} mice.

Specifically, DBA mice may avoid alcohol on the basis of preabsorptive (*e.g.*, taste and olfactory) cues, rather than on the pharmacological consequences of ethanol. C_{57}BL mice, on the other hand, may be less sensitive to these preabsorptive cues, and thus postabsorptive cues may be more relevant in explaining alcohol preference in mice of this genotype.

Thus, it might be of interest to see whether a specific subline of C_{57}BL mice (C_{57}BL/Uae), which have been reported to avoid 10per cent ethanol solutions, show DBA like responses to the preabsorptive cures of ethanol demonstrated by Belknap and associates.

Peripheral Effects

Although drug action on Behavioural phenotypes is often assumed to occur via effects on the central nervous system, it should be apparent from the preceding that genetic influences on these drug responses may operate through peripheral mechanisms. Dispositional and pharmacokinetic factors are, however, not the only characters of this type.

Cardiovascular as well as autonomic nervous system effects of pharmacological agents are well known. It is conceivable that alteration of such characters as heart rate, blood pressure, and sympathetic tone may play a large role in the general Behavioural response to a drug. In other words, subjective evaluation of drug response may depend on information about changes in activity of these peripheral systems.

A probable genetic component in variability of peripheral response to alcohol is well known. Racial differences in such variables as heart rate response, blood pressure effects and facial flush response are documented. The greater sensitivity of Orientals to alcohol may be due in part to metabolic factors. However, the marked differences in peripheral response to alcohol may be related to differential drinking patterns as well.

The issue may be even more complex, as suggested by Zeiner, Peredes, and Christensen, who investigated the role of the primary metabolite of ethanol, acetaldehyde, in these racial differences in peripheral response. If acetaldehyde is responsible for ethanol effects, then examination of its direct effects is required.

CENTRAL NERVOUS SYSTEM

Within the central nervous system (CNS) there are a number of physiological/ biochemical loci where genetic influences and drug effects might converge. One important way in which such genetic influences might occur. Genetic control over CNS development and the general issue of epigenetic processes. There remain several ways in which the genome can exert an influence over functional states of the CNS and, both by demonstration and inference, Behaviour.

Membranes and Channels

The unique attribute of the neuron, its excitable membrane, depends on the regulated passage of several kinds of ions through specialized protein channels. The action potential is in reality the generation of sodium and potassium currents as they traverse the neural membrane through these channels. The release of several neurotransmitters is dependent on similar actions of calcium.

Modification of the number of these channels or their willingness to pass respective ions would affect neural function at its most basic level. Genetic control of affinity might be direct, given a single amino acid change due to mutation. Channel number control would seem to depend on gene regulation or posttranscriptional/epigenetic influences.

Several channel mutations have been isolated and mapped in paramecium species. Although this is not an example of a neural phenomenon, it remains an impressive fact that calcium and potassium channel mutants have Behavioural sequelae in these single celled organisms. A particularly interesting use of these mutants was reported by Schein, who took advantage of asexual reproduction and induced autogamy to follow alterations in membrane excitability during changes of paramecia from heterozygote to homozygous mutant form.

Such a procedure allowed estimation of the calcium channel half life at 5-8 days. Although no such demonstrations are, to our knowledge, available in neural tissue, the importance of such a possibility cannot be overlooked.

It has been long known that the narcotic effectiveness of depressant agents such as alcohol is related to the degree of lipid solubility. Thus the degree to which a compound can distribute itself into neural membranes is a direct predictor of its anesthetic potency. Because it is well accepted that anesthetics and many sedative hypnotic drugs can act in this way, it is a parsimonious assumption that genetic variation in membrane construction could result in differing lipid solubility of the same drug in different genotypes.

Although such genetic variation has, to our knowledge, not yet been demonstrated, the demonstrable hereditary variation in neurosensitivity to alcohols seems likely to involve these membrane mechanisms. The direct action of anesthetics on membranes is thought to occur via alteration of the proteins found there; these proteins are likely to be those of the channels.

One investigation in this area has attempted to use genetic variation as a tool for studies of possible commonality of action of ethanol and halothane, a gaseous anesthetic. The LS and SS mice described earlier were shown to be similarly affected by halothane, even though their ethanol response is so strikingly different.

It was concluded that the two drugs act via different mechanisms. In the present context, it would appear that different classes of membrane proteins are being affected by the two drugs. Alternatively, the differential LS and SS response to ethanol may not take its origin in membrane effects.

Receptors

In recent years it has become increasingly apparent that the functioning of the brain is critically dependent on highly specialized receptor molecules, protein in nature, and thus potentially subject to genetically controlled variation. These receptors are usually thought of as specific for particular neurotransmitters, neuromodulators, or hormones. Several clear examples of the involvement of such receptors exist in the psychopharmacogenetic literature. We deal here with drug effects and receptor studies of three types: benzodiazepines, opiates, and barbiturates. Benzodiazepines have recently been shown to bind to specific sties in CNS membrane fragments. Thus it appears that the nervous system may possess endogenous neurotransmitter systems that are similar in function to the functions affected by this class of tranquilizers. The clinical efficacy of benzodiazepines (*i.e.*, their anxiolytic effects) tends to correlate well with their affinity for these specific receptors. It has been suggested that characteristics of such endogenous benzodiazepine systems might be altered in pathological disorders of affect that are usually treated with this drug class. In animal model research, demonstrable genetic variation in emotionality/reactivity seems a likely candidate for explanation by variation in these receptors. Towards this end, Robertson, Martin, and Candy have studied benzodiazepine receptors in the rat strains selectively bred by Broadhurst for reactivity and nonreactivity.

The Maudsley Non- reactive rats (MNR) had a higher specific binding of diazepam in every brain region examined than did the Maudsley reactive (MR) rats. The largest differences were found in limbic structures.

The nature of the differences was primarily one of receptor number rather than affinity. In hypothalamic tissue the number of binding sites was 1000 fmol/mg protein in MR rats and 1385 fmol/mg protein in MNR rats. The affinity for the ligand was only slightly higher in the MNR strain: $K_D = 3.64$ nmol for MR and 3.24 nmol for MNR. These authors suggest this issue of receptor density as a plausible basis for the reactivity differences in the strains.

Presumably the more reactive MR rats have a less extensive endogenous "tranquilizing" system. In another study from the same labouratory similar results were found in four mouse strains. "Emotional" BALB/c mice had a lower density of benzodiazepine receptors than did three "less emotional" strains. Both these studies reinforce the notion that endogenous benzodiazepine systems play a role in emotional Behaviour. A more complete analysis in the future will, we hope, combine this type of study with actual examination of benzodiazepine effects on "emotionality."

Another type of receptor, the opiate receptor, has been similarly studied. Considerable genetic diversity has been demonstrated for Behavioural effects of opiate compounds in mice. Again, a parsimonious prediction might be that opiate receptor density or affinity could serve as the basis for varied Behavioural effects of opiates. Two groups of investigators have directly addressed this issue.

Baran et al. investigated naloxone binding and analgesic response to morphine in C_{57}BL/6By, BALB/cBy, their reciprocal F_1 hybrids, and seven recombinant inbred strains derived from those parental stocks. Considerable strain differences for total stereospecific naloxone binding in whole brain and also analgesic response were reported. Even though the strain with the lowest naloxone binding (CxBK) also had the lowest analgesic response, the correlation between binding and analgesia across strains was a Non- significant.48. The relatively low statistical power for this correlation suggests that a real, but minor relationship may exist. Crabbe and Belknap point out that genotypic variation in Behavioural response may well arise in part from dispositional factors, and thus the relative importance of receptor characteristics would diminish.

We would suggest two additional possibilities. First, the Baran et al. study did not examine specific brain regions. To the extent that analgesic effectiveness of morphine is mediated by specific brain regions (*e.g.*, the periaqueductal grey), analysis of a heterogeneous whole brain could well obscure important predictability from specific regions.

The notion of Non-homogeneous opiate receptor system distributions raises a second issue. Present conceptions of opiate systems within the brain assume not only anatomical topographies, but biochemical heterogeneity of opiate receptor subtypes. It is conceivable then, in this context, that naloxone specific binding may not reflect a more specific receptor subtype, which holds importance for the genetic variation in analgesic potency. Use of another ligand might reveal a correlation with the genetic effects on analgesia.

A recent study has examined the role of subpopulations of opiate receptors in genotype influenced morphine sensitivity. This work with C_{57}BL/6 and DBA/2 strains of mice found that strain differences in opiate receptor number depended on both ligand (and thus subpopulation) and brain region. The C_{57}BL strain had considerably more striatal opiate receptors when D-Ala$_2$ -Met-enkephalin or Leu-enkephalin were used as ligands. Such differences did not occur when naloxone or di hydromorphine were ligands, nor did they occur in three other regions examined.

These results suggested that the locomotor response to morphine in C_{57}BL mice could take its origin in this striatal subpopulation of opiate receptors. It is interesting that neither of these studies on strain differences in opiate receptor characteristics reported differences among genotypes in affinity (K_D) of the ligand for the receptor.

This suggests that the kind of genetic variation present may not be that of structural gene mutation but rather complex regulation of the number of receptors inserted into neural membranes, or perhaps cell number. In this context, it may well be that control of receptor number in neural tissue could serve as a prime research area for regulation of gene action in higher organisms. A developmental perspective on this issue would probably also be of value. As a third example of receptor characteristics and genetic control of drug response, the work of Waddingham, Riffee, Belknap, and Sheppard is of interest. The regulation of membrane bound â-adrenergic receptors by catecholamines and the ubiquitous nature of norepinephrine influences on a variety of Behaviours suggest these receptors to be of import in neural function. This study examined genetic influence on these â receptors and their response to phenobarbital treatment. C_{57}BL/6 mice had generally higher densities of â adrenergic receptors in several brain regions. After phenobarbital treatment, receptor proliferation was apparent, but dependent both qualitatively and quantitatively on strain and brain region. Again here, regulatory gene activity may be of interest particularly in the framework of an inducible system—induction produced by phenobarbital treatment.

ENZYMES

As major gene products, enzymes in the CNS are likely candidates for heritable variation. The range of enzymes active in neural tissue is wide and provides a wealth of material for pharmacological study. At a time when research on Behavioural control by various neurotransmitter systems is so predominant, it is perhaps easy to overlook other kinds of enzyme systems that could markedly influence drug response. Not the least of these are those biochemical pathways involved in energy metabolism.

Small perturbations in cerebral energy metabolism can have potentially major effects on a wide variety of Behaviours. The example provided by hypoglycemia and its effects on cognitive and affective characters is clear. The brain is a high energy utilization organ. All specialized functions of neural tissue depend on the more basic biochemistry of energy sources. Although we are not aware of genetic studies of enzymes involved in such energy metabolism and pharmacological alterations of activity, the caveat is relevant in the context of the distinction between direct and indirect drug effects on Behaviour.

A drug such as alcohol may very well modify neurotransmitter activity and some related Behaviour by actions on synthesis, degradation, receptor binding characteristics, or channel alteration. Nu merous other "side" effects might also occur. For example, ethanol consumption produces a marked alteration in general metabolism, producing among other things a general acidosis and increasing blood a-hydroxybutyrate levels. As â hydroxybutyrate is a usable energy source for brain tissue, marked changes in cerebral energy metabolism are likely to follow.

Such "indirect" effects on neural function, and perhaps Behaviour, would certainly appear subject to the same sort of heritable variation as other biochemical systems. Thus the degree of Behavioural modification by indirect as well as direct action of the pharmacological agent might also be influenced by genetic factors. While hoping not to be redundant, we believe the conclusion is obivous. That is, gene action may affect pharamcological responsiveness in a variety of ways other than via the specific neural system controlling the phenotype of interest. Keeping the foregoing caution in mind, numerous informative avenues of research on neuro-transmitter system regulation by genetic variables are available. The most well researched of these is probably the area of catecholamine neurotransmitters. Norepinephrine and dopamine systems are subject to functional alteration by numerous pharmacological agents; many of these agents are of use in a clinical setting precisely because of their Behavioural effects.

These functional alterations can occur as a result of inhibition or induction of the enzymes responsible for synthesis and degradation of these catecholamine transmitters. Early work by Ciaranello, Barchas, Kessler, and Barchas first established the fact of strain differences in brain activity of tyrosine hydroxylase (TH), the rate limiting enzyme in catecholamine biosynthesis.

The same report also demonstrated marked strain differences in adrenal phenylethanolamine-N methyl transferase (PNMT), the enzyme responsible for epinephrine production from norepinephrine. We here describe further examination of these mouse strain differences in TH activity in a context that serves as an example for pharmacogenetic study as well as neural development.

In midbrain of CBA/J mice, the activity of TH is 20per cent less than in BALB/cJ mice. A comparable difference in immunocyto chemically detectable dopamine neurons in midbrain suggests that this difference in TH activity is due to more cells with the enzyme, rather than to greater specific activity per cell. A correspondingly more dense innervation of the striatum, by what are presumably the axons of these midbrain nigral cells, occurs in the BALB/c mice.

It is likely that this difference is related in a general way to the fact that the BALB/c brain is simply a larger one. Non-etheless, this represents a clear example of genetic control over neuronal organization and neurochemical characteristics of a Behaviourally important system. The differences in density of striatal innervation apparently predispose a differential response to two agents with dopaminergic activity, amphetamine and apomorphine.

BALB/c mice are more responsive to the locomotor stimulant effects of amphetamine and less sensitive to the stereotypy producing effects of apomorphine, a dopamine receptor agonist. Interestingly, these charac teristics follow a developmental time course that lags slightly behind the appearance of the TH activity difference. Apparently both pre and postsynaptic functions of the dopamine system are different between the two strains as reflected in the actions of the two drugs.

A more detailed understanding of the genetic control over dopamine cell proliferation, growth, and effects on target tissue in this model system will answer many questions not only of pharmacogenetic interest, but also of basic importance to developmental neurobiology.

In this example of TH activity differences, the enzyme activity itself is probably not the focal point of either gene or pharmacological action. As noted, a more basic developmental issue is central. Yet the enzyme is a clear marker for a dramatic genetic influence. In other work, TH is seen to be more directly involved.

Gamma-butyrolactone (GBL), through its metabolite gamma-hydroxy butyric acid (GHB), produces an inhibition of impulse transmission in dopamine neurons. A compensatory activation of TH occurs in these cells and an accumulation of dopamine occurs. In the LS and SS mice, GLB has hypnotic effects that differ in the same direction as ethanol effects. Of interest here is the fact that dopamine accumulation is 50per cent greater in the LS mice after GBL treatment. Presumably the activation of TH was greater in LS mice. This activation occurs via kinetic rather than molecular genetic induction and apparently reflects a differential effectiveness of GHB at a subcellular site in LS and SS dopamine neurons.

Although several studies of neurotransmitter systems' enzymes exist, there is yet to be a complete characterization of all biosynthetic and degradatory enzymes, transmitter levels and turnover rate, and receptor kinetics for any single system where genetic variation is also of interest. The difficulties of such a massive undertaking are obvious, but the benefits accruing to all three fields, neuropharmacology of drug action, neurochemical organization and dynamics, and mechanisms of genetic action are equally impressive. The field must surely progress along these lines.

ALCOHOL

The genetic effects on a given Behavioural response do not necessarily imply similar genetic effects on other phenotypes. The degree to which a similar genetic pattern occurs for more than one phenotype is a measure of the genetic correlation of these phenotypes. Yet taskor situation specific genetic effects are not rare. The LS and SS mice are usually very different in their response to ethanol.

Yet in one study, across a wide range of doses, a particular Behaviour was not differentially affected. When ethanol was used as a poison (UCS) in a conditioned taste aversion experiment, the magnitude of aversion produced was similar in the two lines. Thus, whatever systems underlie the loss of righting relfex differences do not control response to ethanol in the taste aversion experiment.

A second example may be drawn from the literature on strain differences in ethanol response. Although BALB/c mice lose the righting reflex longer than C_{57}BL/6 mice, they show a greater locomootor stimulation following low dose ethanol administration. It may be of interest that, in these two strains, amphetamine shows exactly opposite effects on locomotor activity. BALB/c mice are not activated, whereas C_{57}BL/10 mice are thus a negative genetic correlation is suggested for these effects of amphetamine and ethanol.

MEASURES OF RESPONSES TO OPIATES

It is becoming increasingly apparent that strain differences are readily obtained using the standard measures of opiate induced analgesia and locomotor activity. However, it is equally apparent that both the magnitude and the direction of these differences change with the particular measures and procedures employed. For example, C_{57}BL mice have been characterized as being sensitive to morphine induced hyperactivity in open field tests and toggle boxes.

However, Eidelberg, Erspamer, Kreinick, and Harris have reported that C_{57}BL mice respond to several doses of morphine with decreased activity when the measure is recorded on an activity wheel. Similarly, measures of analgesia, such as the hot plate technique, tail flick procedures, and abdominal writhing induced by agents such as acetic acid often yield very different strain distribution patterns. The exact nature of these differences among measures has been reviewed elsewhere.

For the present, the disparate results of these procedures lead us to argue for careful descriptions of the methods employed when general phenomena associated with opiate responsivity. However, we share the optimism of Crabbe and Belknap that consistent patterns among several dimensions of opiate responsiveness may yet emerge as the results of further investigations are known.

GENETIC CORRELATIONS

The basis of many inferences in pharmacogenetic research is the comparison of effects of more than one drug on the same genetic material. This is implicitly an example of genetic correlation. Both correlated drug effects and correlated effects of a single drug on several Behaviours are means for analyzing the domain of effects of that pharmacological agent, or the common physiological substrate of a variety of agents.

The quantitative genetic methods for deriving genetic correlations are readily available and are not presented in detail here. Instead, the major experimental approaches to genetic correlation methods and the general usefulness of such correlations.

One of the most utilized approaches to genetic correlations has been the examination of correlated characters in selected lines. A generalized example would find a set of lines selected for response to drug X, characterized at a variety of physiological levels, and perhaps other types of responses to the same drug being assessed as well. The usual argument is that care must be taken in interpretation of any correlated characters especially if replicated lines and unselected control lines are not available.

The basis of this assertion is that when only two lines are sampled and two characters are measured on those genotypes, then the degrees of freedom for assessing the correlation are $N - 2 = 0$. Two points are worthwhile in counter. A set of once replicated lines would produce a situation with $df = 2$, hardly a powerful statistical test.Only when the numbers of genotypes exceeds 20 will actual correlations below.40 be detectable. In addition, correlational degrees of freedom are based on the assumption that both variables generating the covariance are free to vary. This is clearly not the case after a selection has been accomplished.

The selected phenotype is at that point fixed. Only the as yet unmeasured phenotype is "free" to vary. Perhaps a more Bayesian approach is desirable.The most rational alternative to numerous replicates is a simultaneous measurement of several phenotypes during the course of selection. As selection progresses, those characters sharing a true genetic covariance with the selected phenotype will diverge in somewhat parallel fashion. As this approach forces one to identify relevant variables even before the selection begins, it is a less than perfect one.

Obviously a more practical approach is required. We would advocate a rational approach that makes use of a priori knowledge of the pharmacological system being examined and physiological intervention if the question being asked is one of mechanism of drug action. For example, in the LS and SS mice it is not surprising that alcohols other than ethanol have a differential effect.

However, if a correlated character is, for example, an altered preference for saccharin sweetened water, then one must wonder if drift is responsible. In sum, no single approach seems completely suitable. Therefore, only careful and objective scrutiny can avoid erroneous conclusions. An additional tool in genetic correlation analysis is the diallel cross and also the triple test cross. The inclusion of Non- homozygous animals in the diallel cross analysis is probably an advantage, but again, unless the number of genotypes is large, statistical power is low.

The use of genetic correlations to assess commonalities of drug action is a powerful approach. Only the use of cross tolerance studies can so clearly demon strate common substrates of action as can the presence of a genetic correlation. The aforementioned effects of higher order alcohols on LS and SS mice led to the conclusion that considerably different mechanisms of action underlie the effects of alcohols and other sedative hypnotics that did not differentially affect the two lines.

It is conceivable that, in a genetic system with high heritabilities, an experimenter could test a very large range of drug classes and attempt to classify them via one of a number of classification techniques on the basis of the genetic correlations.

The alternate approach examines the effects of a single drug on numerous Behaviours. Examination of the kinds of genotype patterns of Behaviours affected similarly by a given drug should indicate similar substrates for those Behaviours. It would also be of interest to examine the pattern of genetic correlations in drugged and undrugged mice separately. Changes in those correlations would suggest an ability of the drug to dissociate genetically based covariation of Behaviours. Thus even at a strictly Behavioural level of analysis, the genetic correlation could be a very useful tool.

DEVELOPMENTAL PHENOMENA

As already mentioned, the examination of drug responses in the developing organism and concomitant study of neural development is a powerful approach to the study of drug action and neural development. The addition of the genetic variable as a unique tool to tease apart neuropharmacological mechanisms is a potentially fruitful tactic. Correlated biochemical and pharmacological changes are more easily linked if the correlation holds in genotypes that show different time courses of developmental events. An example from the literature on alcohol preference in mice is instructive. Wood reviews this literature on age related differences in response to and preference for alcohol by labouratory rodents. Two studies in particular are instructive both for purposes of the genetic variable and its interaction with an experiential variable. BALB/c mice are, as adults, avoiders of ethanol.

Yet, from 4-9 weeks of age, they will voluntarily consume a fair amount of ethanol. Consumption diminished to near zero by 9-10 weeks unless a choice between alcohol and water was provided during the 4 to 8-week period. Interestingly, mice forced to drink alcohol during this period would typically not consume the alcohol as adults. Thus the fact of choice in young mice appears to predispose alcohol consumption in adults. This plasticity of phenotype or lack of phenostability makes clear the complex nature of the neural and Behavioural systems on which the genes can act only very indirectly.

A more general view of pharmacogenetic research must realise that a most critical variable in drug effectiveness is the functional organization of the complex target organ called the nervous system. It is ultimately the genetic influence on the assemblage of the brain and other physiological systems that provides the substrate for individual differences produced in drug responsivity.

A given drug acting on similar receptors in two brains that differ in organization of the neural network in which these receptors reside might produce different Behavioural outcomes. The mechanism of such gene effects on neural development may be as simple as cell proliferation and migration rates or as complex as regulation of the timing of critical and sensitive periods for Behavioural and emotional development.The study of such biochemical characters as those already enumerated may yield important findings. Ultimately, however, a concern with developmental issues in pharmacogenetic work will yield much more information about the mechanisms of gene and drug action.

9

Genetics Disorders

PHYCHOLOGICAL DISORDER

Schizophrenia is one of the more complex and less understood of the psychological disorders. For many years, studies have explored the genetic aspect of this syndrome in an effort to provide a better understanding of its etiology, symptoms, course, and development. Much of what is involved in schizophrenia still remains unclear, although great strides have been accomplished by research in the area.In trying to understand the etiology of psychiatric disorders, one cannot avoid studying gene—environment interactions. Just as genes influence the susceptibility of an individual to the development of almost any disorder, environmental factors are also seen as important etiological contributors to disturbance. Consequently, whether or not genetic or environmental variation is involved in the development of schizophrenia is no longer considered to be an important point of contention in this area of research. Instead, most of the current investigations appear to be concerned with acquiring a more thorough understanding of the magnitude and relevance of both environmental and genetic variation and the ways in which these interact to establish the various manifestations of schizophrenia.

Today there seems to be little doubt that a genetic basis for schizophrenia exists. However, the specific schizophrenia-producing genetic components and the method by which they are transmitted remain unclear. Genetic investigations have involved primarily three methodologies: family studies, twin studies, and adoption studies. Despite some difficulties, these methods have obtained strong evidence pointing Towards the involvement of genetic factors in the development of schizophrenia. Currently, the amount of interest and speculation in this area of research is high, and on the basis of findings, sophisticated models and new, more innovative research techniques are being developed. The major conceptual, diagnostic, and methodological difficulties inherent in this area of investigation, the evidence for the existence of genetic factors in the etiology of schizophrenia, and the models for a genetic mode of transmission that have arisen from these lines of research.

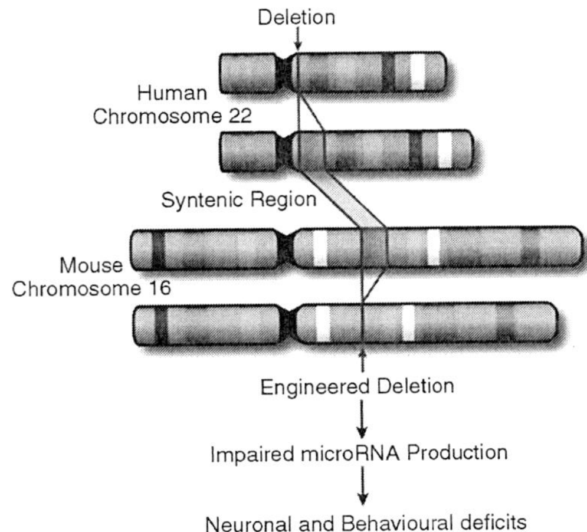

Fig. Genetics of Schizophrenia

One of the more difficult aspects of this field remains the lack of knowledge concerning the nature of schizophrenia. To date, no homogeneous group of schizophrenics has been isolated and no unitary etiology for the disorder has been found.

The possibility exists that schizophrenia is not a single, unitary disorder, but may consist of a number of different entities. Schizophrenic heterogeneity could be present at several levels. Seemingly similar phenotypic expressions may be due to quite different etiologies. As a result, the reliance upon symptomatology for the differentiation of schizophrenic subtypes may be grossly misleading. To complicate matters further, the complexity of gene operations makes it difficult to delineate the various interacting etiological components of a schizophrenic disorder.

The nature of the activating environment and the timing of that activation in addition to the gene could all combine to produce a single phenotype. Therefore, a range of possible phenotypic outcomes exists for every gene as a function of the activating environment and the time of activation. Depending on the environmental circumstances, there are a number of different genes that can produce a single phenotype. Consequently, several genetic mechanisms may underlie the expression of a single characteristic.

If schizophrenia is in fact heterogeneous, many of the present research strategies may prove inadequate because most rely on the implicit assumption of homogeneity among patients studied. A schizophrenic sample based on an overinclusive diagnostic category would result in variable schizophrenic performance. As a result, common findings among schizophrenics would be difficult to detect and valid hypotheses may be prematurely discarded on the basis that the data fail to meet statistical significance.The failure to reach a clear consensus on the nature of schizophrenia has resulted in conflicting viewpoints. To date, genetic studies of schizophrenia have relied almost exclusively on the assumption of a model in which schizophrenia consists of a complex disorder amenable to the same type of study and displaying the same pathogenic characteristics as an inheritable disease.

Other points of view, especially those among the psychodynamic and contextualist schools of thought, hold that the conceptualization of schizophrenia as a disease entity is inadequate and unjustified. For example, Sarbin and Mancuso claim that schizophrenia is a label and the product of social processes resulting in the imposition of a moral verdict in response to certain undesirable Behaviours. Others have proposed that certain forms of familial interaction predispose an individual to schizophrenia, whereas still others have preferred to focus on the schizophrenic's intrapsychic world and dynamics. An interesting model postulated by Meehl incorporates both genetic and environmental factors to explain the occurrence of schizophrenia. Meehl suggests that all that is inherited in schizophrenia is an integrative neural defect that he labels *schizotaxia*.

The imposition of a social learning history on schizotaxic individuals results in a schizotypic personality organization. That is, all schizotaxic individuals obtain a schizotypic personality organization regardless of environmental conditions. Most of these individuals will remain compensated. However, those who encounter certain detrimental influences (*e.g.*, schizophrenogenic mothering) will decompensate into schizophrenics. In light of this model, only a few of those individuals who have inherited the neural defect and are thereby predisposed to schizophrenia, will manifest the disorder.

Despite the wide disparity among theoretical models, most can be classified as elabourate extensions of the nature versus nurture controversy. It is apparent that this debate cannot be adequately resolved until a broader base of knowledge is developed.

Until that time, it may be wise to bear in mind that genetic research does not necessarily negate endeavors supporting contextualist and psychodynamic conceptions. Instead, genetic research addresses both environmental and genetic factors, and is interested in delineating the effects of each separately and in interaction as they pertain to schizophrenia.

DIAGNOSIS

Another research difficulty related to that of schizophrenic heterogeneity pertains to clinical diagnosis. The new DSM-III allows for considerable heterogeneity as to who will be diagnosed schizophrenic; persons need to fulfill only a certain number of criteria out of a standard set of possibilities in order to receive the diagnosis.Schizophrenia cannot be diagnosed unless signs of the disorder have been present for at least 6 months. Illnesses of shorter duration, or illnesses that have resulted in the face of a psychosocial stressor and appear to be transient in nature are no longer included.In fact, these disorders are not even considered to be subtypes of schizophrenia. These changes in diagnostic criteria have left schizophrenia researchers in a peculiar bind since they now must try to define the onset of a schizophrenic disorder. In the past, the onset of schizophrenia was demarcated by the appearance of psychotic symptoms. However, it is well known that the onset of schizophrenia is often quite insidious, and that prodromal features commonly predate psychotic symptoms.

Logically, it would be best if we could define onset as the point in time when one could predict with virtual certainty that a schizophrenic psychosis would occur. Unfortunately, we still lack the necessary knowledge to make such a prediction, and the definition of onset remains strictly arbitrary. Consequently, the conceptual problem of trying to determine criteria for delineating signs along a continuum marking transitions from a schizophrenia-prone personality structure to a prodomal syndrome to a full-blown psychosis still exists.

Because of these diagnostic difficulties, a standard definition for what constitutes schizophrenia remains to be determined. The absence of a standard definition remains a major hindrance to the progression of this field. The severity of the problem is due to the fact that every research project must of necessity concern itself with identifying those people who have the disorder.

Therefore, the homogeneity of the sample studied and the Rigour of the investigation carried out on the sample depend on the accuracy of the diagnosis. Unfortunately, there is no independent validation and only moderate reliability for diagnosing a schizophrenic disorder. Even the combination of moderate interrater reliability and independent validation appears to be inadequate. Consequently, there is no truly homogeneous sample in schizophrenia research. Instead, investigators are forced to contend with the heterogeneous groupings that fit the schizophrenic label, the abundance of contradictory data, and the numerous disagreements over research findings. As long as this state of affairs exists, studies are limited to statistical rather than clinical significance.

At present, many diagnostic difficulties result because little is known about the unity or diversity of schizophrenia. In order to gain a clearer understanding of the nature of the disorder, many investigations have been conducted to determine the degree to which the range of phenotypic expression in schizophrenia is paralleled by genetic heterogeneity. These included certain studies that were designed to investigate whether these subtypes were fundamentally alike or different.One such study by Winokur, Morrison, Clancy, and Crowe found that hebephrenic probands had more than three times as many schizophrenic relatives as paranoid probands. These authors determined that both children and other secondary cases of schizophrenia in the families of hebephrenics were more likely to be diagnosed hebephrenic as opposed to paranoid schizophrenic. The same type of pattern was established for paranoid schizophrenics.The children and other secondary cases of schizophrenia in the paranoid probands were more likely to manifest paranoid rather than hebephrenic symptoms. These familial data led the au thors to conclude that process schizophrenia is not likely to be a unitary illness. Instead, they postulated the existence of two types of schizophrenia in the population studied, one of which may manifest itself mostly as hebephrenia but also occasionally as paranoid schizophrenia.

The second type would manifest itself exclusively as paranoid schizophrenia in which there are relatively few schizophrenic relatives. Together, these findings suggest that there may be only one type of hebephrenic schizophrenia, but two types of paranoid schizophrenia, one related genetically to the hebephrenic subtype.

Another study by Kety et al. in 1975 demonstrated that the diagnoses of chronic, latent or borderline, and uncertain schizophrenia tend to cluster in the biological relatives of schizophrenic adoptees. Similarly, a study by Slater and Cowie examined the classical pre-Kraepelinian categories and found a significant association between schizophrenic probands and their schizophrenic offspring with respect to type of category.

Kringlen in 1968 corroborated these results and also demonstrated that the entire set of Kraepelinian subtypes is represented in the children of every type of schizophrenic proband.

These studies are typical of investigations in this area. Their results seem to indicate that some groups of disorders are genetically related, whereas the relationships among others remain unclear. Despite some contradictory findings, the most common results from these studies provide evidence for a common genetic basis among various forms of schizophrenic disorders.

In general, the findings indicate that cases of all subtypes occur in the families of each schizophrenic subtype. They also suggest that only a modest tendency Towards subtype resemblance exists within families.

In addition to considering the unity or diversity of schizophrenia, one must also consider the relationship between schizophrenia and other psychiatric disturbances when making clinical diagnoses. This is especially important if one wants to formulate diagnostic criteria that are successful at differentiating among various psychoses.At present little is known about the specificity of the schizophrenic disorder. However, there is some evidence suggesting a genetic relationship between schizophrenia and other psychiatric disorders. In a study by Reed, Hartley, Anderson, Phillips, and Johnson, a high risk of psychoses for the siblings of psychotics was found.

When the investigators looked at specific disorders, they discovered that the risk for first degree relatives of affective probands to develop a similar disorder was less than their risk of becoming schizophrenic. There is also some indication that the risk of schizophrenia for the offspring of paranoid and hebephrenic schizophrenics is lower than the risk for children of manic—depressives. Even some of the very earliest studies discovered that schizophrenic and affectively ill individuals were about equally prevalent in the families of schizophrenics. Today as well, despite efforts to select clinically homoge neous probands, the occurrence of different types of psychoses in the same family is difficult to avoid.

The suggestion that there may be a similar genetic basis to schizophrenia and other forms of psychoses raises the possibility of overlap in the symptomatology of various psychiatric disturbances; and indeed, this has been found to be the case. Another problem exists in the possibility that symptomatology is not directly related to genetic inheritance. Bleuler speculated that secondary symptoms such as delusions and hallucinations are not inherited. He also suggested that the primary symptoms do not correspond to the transmitted anomaly itself.

As a result of these possibilies, there are some major difficulties involved in basing diagnostic criteria on clinical symptoms. However, there are now some indications that future diagnostic categories will be based on biochemical indicators rather than Behavioural symptoms in order to decrease the heterogeneity of individuals in specified subgroups of psychopathology.

Using biological markers could be advantageous in several ways. First, it could avoid the genetically heterogeneous diagnostic groupings based on standard methods of symptom classification. It is well known that even relatively well formulated clinical subtypes of schizophrenia present diverse symptomatology.

As a result, it is impossible to differentiate between genotypic and phenotypic variations under the present diagnostic systems. The use of biologically independent variables in formulating diagnoses may promote the detection of biological etiologies and the formation of more homogeneous psychiatric diagnostic categories. Individuals who exhibit divergent or less severe clinical symptoms and are thereby excluded from present investigations could be included if they met a biological criterion.This in turn would allow the high risk population (those individuals with the salient factor) to be identified and studied. The use of a biologically based strategy would also permit the study of Non- hospitalized and unmedicated individuals, thereby avoiding the inclusion of these confounds within investigations.

In addition to developing biologically based criteria, many other efforts are being made to improve the diagnostic system. For example, the United States—United Kingdom Diagnostic Project, the International Pilot Study of Schizophrenia, and the World Health Organization have provided standardized input on patients for formulating diagnoses, experts to address diagnostic issues, and cross national reference points in order to establish criteria that will improve the reliability and validity of the diagnosis of schizophrenia on an international basis.Additional reasons for optimism stem from the new trends in schizophrenia nosology. These include an increased reliance on affective symptoms as diagnostically useful, as well as a shift Towards incorporating the schizo affective disorders within the affective diagnoses. In addition, the broad use of the schizophrenic label seems to be decreasing as a function of less emphasis being placed on classic schizophrenic symptoms, and more effort being devoted to the use of operationally defined criteria tested for reliability.

Although at present it is still difficult to reach clear and definite conclusions in research dealing with schizophrenia because of the many diagnostic difficulties associated with this disorder, the increasing refinement of diagnostic issues and the development of more sophisticated research techniques makes it possible to conceive of removing the current obstacles to gathering clinically significant data and conducting therapeutic—preventive research.

FAMILY STUDIES

Despite the numerous conceptual and diagnostic difficulties inherent in schizophrenia research, much evidence has been gathered to substantiate the existence of genetic components in the transmission of the disorder. The idea that various forms of mental illness are heritable has existed for many centuries. However, it was not until early in this century that researchers were able to provide actual documentation of this fact.

The first consanguinity study that found a significant increase in the prevalence rate of schizophrenia among relatives of index cases in comparison to the rate of the general population. More specifically, he discovered an age corrected risk for schizophrenia of 4.48per cent and 4.12per cent for other psychoses in the siblings of schizophrenics whose parents were unaffected.

Many other studies have confirmed Rudin's basic finding of a higher rate of schizophrenia in the relatives of samples of schizophrenics than in the samples from the general population. From these studies the rate of schizophrenia in the general population has been calculated to range from.35 to 2.85per cent, with a mean around.85per cent.

The findings from the consanguinity studies also reveal a clear correlation between the degree of blood relationship in relatives of an individual affected with schizophrenia and the risk of those relatives for becoming schizophrenic. More specifically, those relatives who share fewer genes in common with an affected individual show a lower expectancy rate for the disorder than those who share more genes.

In addition, these studies have demonstrated that there are no significa nt differences between prevalence rates for siblings and other first degree relatives. Although the difference found in the rate of schizophrenia between parents (5.5 per cent) and siblings and children of schizophrenic probands (10.2per cent and 13.9per cent respectively) appears to be large and therefore contradictory to the foregoing statement, this difference has been explained by the fact that these parents are largely selected for a degree of mental health through having been married and having children.

When a correction is made for this factor, the risk in parents of schizophrenics becomes close to that of siblings and children. The somewhat higher risk in children than in siblings of schizophrenics is thought to be due to sampling error. Additional evidence for a genetic factor in the transmission of schizophrenia is found by comparing the different rates of prevalence for the disease among offspring of matings of various psychotic combinations of parents. Estimates have been obtained indicating that the risk for schizophrenia increases as the genetic loading increases. The best estimate of the risk of schizophrenia in the sibling of a schizophrenic is 9.7 per cent when neither parent has the disorder.

This risk increases to 17.2 per cent for the siblings of the proband when one parent also has schizophrenia. When both parents are schizophrenic, the risk to their children ranges between 36.6 per cent and 46.3 per cent. Although it can be argued that this raise in risk from 17.2 per cent to 46.3 per cent for the siblings of a schizophrenic may be due to the extreme environmental conditions created by the mating of two schizophrenics, a study by Kallman reported that a comparable disturbed setting created by the mating of a schizophrenic and a psychopath yielded a sibling risk for schizophrenia of only 15per cent in Gottesman and Shields. The basic findings of these studies have been further substantiated by the results of more recent investigations. For example, a study by Reed et al. reported that the risk to the offspring when neither parent was psychotic increased relative to the number of aunts and uncles who were psychiatrically disturbed.

The difference among estimates of risk is a fact significant for genetic theories. Not only does this difference demonstrate that the risk of mental illness increases as a function of increased genetic loading, but the large amount of variation in the offspring undermines the prediction of environmental theories that virtually all children from dual matings of schizophrenics would be emotionally disturbed.

Another line of evidence for the role of genetic factors in schizophrenia was provided by Manfred Bleuler, who applied the methodology of consanguinity studies to clinical symptomatology and course over time. He discovered that the biological relatives of schizophrenics who also had the disorder tended to have similar ages of onset, presenting clinical pictures, lengths of duration of illness, and outcomes. These findings can be interpreted in either an environmental or genetic fashion. However, a genetic explanation appears to carry more weight because many of the relatives in the study were geographically separated and in some cases did not even know each other. Therefore, the findings of the study indicate that genetic factors play a role in several clinical parametres of schizophrenia.Although these family studies have contributed much Towards the understanding of schizophrenia and have pointed to the role of genetic factors in the development of this disorder, some major difficulties are inherent in their methodology. The earliest family studies did not question the assumption that schizophrenia is heritable and therefore failed to consider the possibility that environmental factors such as family interactions also contribute to the etiology of the disorder.

As a result, these studies provided sources of data on the empirical risk of the occurrence of schizophrenia in different kinds of relatives, but did not elucidate the specific etiology involved in creating those risks. In other words, these studies could indicate if a trait ran in families, but could not specify what aspect of the family determined the trait. Obviously, families share more than genetic similarity, they also share a variety of social, psychological, and cultural experiences. Consequently, the results of these studies can just as validly be interpreted as supporting an environmental position.

Recently, some studies taking an environmental approach have investigated family interactions on the premise that certain conditions of family life can predispose an individual to schizophrenia. The majority of these investigations have focused on deviant role relationships and disordered communication processes among family members.

The findings of these studies generally indicate that deficits in family communication are highly correlated with schizophrenia and high risks for schizophrenia in offspring. In addition, there are some indications that families of schizophrenics contain poorly articulated and rigid role structures in the areas of dominance, autonomy, and sexuality. Despite the success of this type of research, it is unable to establish the occurrence of schizophrenia as a strictly environmental outcome. In fact, the evidence that is presently available tends to suggest that both genetics and family process play a role in the development of schizophrenia. In light of this evidence, the current mode of investigation has changed to incorporate biological, genetic, psychological, and social factors in the context of complex, multifactorial models of causation when approaching family variables in relation to schizophrenia.

TWIN STUDIES

It was not until the advent of twin studies that the role of genetic and environmental factors in schizophrenia could be more clearly defined. The power of the twin method resides in the environmental similarities and genetic differences between monozygotic and dizygotic twins. Both types of twins share the same intrauterine environment simultaneously with their co- twin. However, monozygotic twins share the same fertilized ovum initially, whereas dizygotic twins are no more genetically alike than ordinary siblings. As a result, one can assume the presence of a genetic factor in determining a trait if the concordance rate for that particular trait is significantly higher for monozygotic as opposed to dizygotic twins.The methodology of twin studies also provides certain advantages. For example, through the comparison of intrapair resemblance, it becomes possible to identify subtypes or components of schizophrenia that may be more under genetic or environmental control than others. Also, by noting the variability of abnormal personality traits in the monozygotic co-twins of typical schizophrenics, it becomes feasible to identify schizophrenic "equivalents" or schizotypes.

The comparison of monozygotic twins who differ in outcome will permit the formulation of inference about the relationship between life experiences and outcome. It must be kept in mind that the twin study method relies on certain assumptions. These specify that monozygotic twins as such are not specifically predisposed to schizophrenia, and that the environments of monozygotic twins are not systematically more similar than those of dizygotic twins in such a way as to foster the disorder.

Also, when assessing the outcome of these studies it is important to be aware of the usage of statistical age corrections and their effects on research findings. Age corrections incorporate estimates of a lifetime risk for developing schizophrenia in order to compensate for a current sample while the disorder exhibits a variable age of onset. Although several methods have been developed, none is entirely satisfactory, and all present only varying degrees of approximation to the true concordance rate.

It is essential to realise that the absolute size of the concordance rate is not predictive of the relative importance of the genetic contribution to the trait in question. Concordance rates can be expressed in several fashions. The basic difference in methods depends on whether concordance is expressed in terms of the pair or the individual.

The casewise or proband method refers to the proportion of cases with an affected partner, whereas the pairwise rate considers the proportion of pairs in which both twins are affected. The individual or proband concordance method yields a higher per centage than a pairwise concordance. Therefore, only those studies that use the same method are comparable.

The most influential early twin studies were conducted by Kallman and Slater. Even though the two studies contained important differences and suffered from methodological problems, both found a significantly higher concordance rate for schizophrenia among monozygotic twins.

By the end of Kallman's study and with some diagnostic revision, but without statistical age correction, 59per cent of 174 monozygotic pairs and 9per cent of 517 dizygotic pairs were concordant for what Kallman labeled "definite schizophrenia." In Slater's study, 65per cent of 37 monozygotic pairs and 14per cent of 58 same sexed dizygotic pairs were found to be concordant. Both studies used the pairwise method for calculating the concordance rates. Findings from other earlier studies were basically consistent with these results. More recently, several studies have been conducted that are methodologically powerful, and that provide highly significant results. One of these studies, reported by Tienari, was conducted on an exclusively male sample in Finland. Initially, this investigator found a zero concordance rate for monozygotic twins.

However, by 1976 the pairwise concordance rate for the sample had risen to 16 per cent. The use of the proband method, the concordance rate for the sample was found to be between 35per cent and 53per cent depending upon which borderline cases were included. As a result, even though it took some time, a significant monozygotic—dizygotic difference was demonstrated.

A study by Kringlen was conducted in Norway and included both male and female twins. His study was especially important for several reasons. His sample was well constructed and contained the largest number of psychotic twins personally investigated among the more recent studies. His investigation was very comprehensive and included case history, pedigree, clinical analysis, rating scale data, and all functional psychoses in addition to schizophrenia. Using the more conservative pairwise method, Kringlen found a significant difference between monozygotic and dizygotic twins with concordance rates ranging from 60per cent for monozygotic probands rated as severe to 25per cent for the most benign.

His results indicated no difference between the concordance rates of monozygotic twins as a function of gender. He reported that whether the sexes of the dizygotic twins were the same or different, the concordance rates did not differ, but were equal to those found in ordinary siblings. These results counter findings from some earlier studies and undermine the psychodynamic thesis that concordance for schizophrenia will increase without regard to genetic similarity, but relative to the amount of likeness and resulting identity confusion among siblings. The findings of another important twin study were reported by Fischer, Harvald, and Hauge in 1969. This study was conducted in Denmark upon a sample that was very representative with respect to twin types and sexes of probands. Only those probands who met strict criteria for chronic or process schizophrenia were included.

Even so, schizophreniform, paranoid, and atypical psychoses in co-twins led to data analysis for quantitative concordance for other than process schizophrenia. As a result of this study, *Fischer et al.* discovered a pairwise concordance rate of 24per cent for monozygotic twins and 10per cent for dizygotic twins.

In addition, the authors calculated the rates according to the proband method and obtained a 36per cent concordance rate for monozygotic and 18per cent concordance rate for dizygotic pairs. When those co-twins who exhibited a form of schizophrenic disorder different from process schizophrenia were included in their calculations, the investigators obtained proband concordance rates of 56per cent and 26per cent for monozygotic and dizygotic twins respectively.These figures appear to be good approximations of the true incidence rate, as the twins used in this study were followed long enough to virtually eliminate the need for any statistical age correction.

Fisher went on to report that monozygotic twin pairs concordant for schizophrenia were not at an increased risk for the disorder when compared to discordant monozygotic twins. She did not discover a relationship between severity of disturbance in the proband and the risk of developing psychosis in the co-twin.

The offspring of concordant twins were not at an increased risk for schizophrenia when compared to the offspring of discordant twins. In addition, the likelihood of a discordant twin member producing a schizophrenic child was equal to that of the schizophrenic member of the twin pair. These findings point Towards a genetic hypothesis and indicate that the genetic transmission of schizophrenia is not always manifested directly.

The Maudsley Hospital based study of Gottesman and Shields conducted in England obtained similar results to those of the studies reported earlier. These investigators also obtained a significantly increased monozygotic concordance rate over that found in dizygotic twins. In a study based on an American sample, found a higher concordance rate for monozygotic as opposed to dizygotic twins. More specifically, the concordance rate for schizophrenia in monozygotic twins was found to be 3.3 times greater than the rate in dizygotic twins.

10

Molecular Genetics

GENOMES

In biology the genome of an organism is the whole hereditary information of an organism that is encoded in the DNA (or, for some viruses, RNA). This includes both the genes and the non-coding sequences. The term was coined in 1920 by Hans Winkler, Professor of Botany at the University of Hamburg, Germany, as a portmanteau of the words gene and chromosome.

More precisely, the genome of an organism is a complete DNA sequence of one set of chromosomes; for example, one of the two sets that a diploid individual carries in every somatic cell.

The term genome can be applied specifically to mean the complete set of nuclear DNA (*i.e.*, the nuclear genome) but can also be applied to organelles that contain their own DNA, as with the mitochondrial genome or the chloroplast genome.

When people say that the genome of a sexually reproducing species has been "sequenced," typically they are referring to a determination of the sequences of one set of autosomes and one of each type of sex chromosome, which together represent both of the possible sexes. Even in species that exist in only one sex, what is described as "a genome sequence" may be a composite from the chromosomes of various individuals. In general use, the phrase genetic makeup is sometimes used conversationally to mean the genome of a particular individual or organism.

The study of the global properties of genomes of related organisms is usually referred to as genomics, which distinguishes it from genetics which generally studies the properties of single genes or groups of genes.

GENOME PROJECTS

Genome projects are scientific endeavours that ultimately aim to determine the complete genome sequence of an organism (be it an animal, a plant, a fungus, a bacterium, an archaean, a protist or a virus). The genome sequence for any organism requires the DNA sequences for each of the chromosomes in an organism to be determined. For bacteria, which usually have just one chromosome, a genome project will aim to map the sequence of that chromosome. Humans, with 22 pairs of chromosomes and 2 sex chromosomes, will require 24 separate chromosome sequences in order to represent the completed genome.The Human Genome Project was a landmark genome project and some have argued that the era of genomics is one of the more fundamental advances in human history.

Genome Assembly

Genome assembly refers to the process of taking a large number of short DNA sequences, all of which were generated by a shotgun sequencing project, and putting them back together to create a representation of the original chromosomes from which the DNA originated. In a shotgun sequencing project, all the DNA from a source (usually a single organism, anything from a bacterium to a mammal) is first fractured into millions of small pieces.

These pieces are then "read" by automated sequencing machines, which can read up to 900 nucleotides or bases at a time. (The four bases are adenine, guanine, cytosine, and thymine, represented as AGCT.) A genome assembly algorithm works by taking all the pieces and aligning them to one another, and detecting all places where two of the short sequences, or reads, overlap. These overlapping reads can be merged together, and the process continues.

Genome assembly is a very difficult computational problem, made more difficult because genomes contain large numbers of identical sequences, known as repeats. These repeats can be thousands of nucleotides long, and some occur in thousands of different locations, especially in the large genomes of plants and animals.

Assembly software

Most research institutes that sequence DNA, use their own software for assembling the sequences that they produce. Some well known assembly programmes include:AMOS (A Modular, Open-Source assembler) is a well-known open source effort to bring together the efforts of leading genome assembly code developers.

The home of AMOS is currently http://amos.sourceforge.net. AMOS was initiated at The Institute for Genomic Research by Steven Salzberg, Mihai Pop, and Art Delcher.The Celera Assembler was the assembler developed by Gene Myers, Granger Sutton, Art Delcher, and others at Celera Genomics from 1998 until approximately 2002.

The Arachne assembler began in 2000 as the doctoral thesis of Serafim Batzoglou, now at Stanford University. Since that time, it has been developed by a team lead by David B. Jaffe at the Broad Institute, formerly part of the Whitehead Institute.

Genome annotation

Genome annotation is the process of attaching biological information to sequences. It consists of two main steps:

1. Identifying elements on the genome, a process called Gene Finding,
2. Attaching biological information to these elements.

Automatic annotation tools try to perform all this by computer analysis, as opposed to manual annotation which involves human expertise. Ideally, these approaches co-exist and complement each other in the same annotation pipeline.

The basic level of annotation is using BLAST for finding similarities, and then annotating genomes based on that. However, nowadays more and more additional information is added to the annotation platform. The additional information allows manual annotators to deconvolute discrepancies between genes that are given the same annotation.

For example, the seed database uses genome context information, similarity scores, experimental data, and integrations of other resources to provide the most accurate genome annotations through their Subsystems approach.

Structural annotation consists in the identification of genomic elements.

- ORFs and their localisation
- gene structure
- coding regions
- location of regulatory motifs

Functional annotation consists in attaching biological information to genomic elements.

- biochemical function
- biological function
- involved regulation and interactions
- expression

These steps may involve both biological experiments and in silico analysis.

A variety of software tools have been developed to permit scientists to view and share genome annotations.

Completion of Genome Projects

When sequencing a genome, there are usually regions that are difficult to sequence (often regions with highly repetitive DNA). Thus, 'completed' genome sequences are rarely ever complete, and terms such as 'working draft' or 'essentially complete' have been used to more accurately describe the status of such genome projects. Even when every base pair of a genome sequence has been determined, there are still likely to be errors present because DNA sequencing is not a completely accurate process. It could also be argued that a complete genome project should include the sequences of mitochondria and (for plants) chloroplasts as these organelles have their own genomes.It is often reported that the goal of sequencing a genome is to obtain information about the complete set of genes in that particular genome sequence. The proportion of a genome that encodes for genes may be very small (particularly in eukaryotes such as humans, where coding DNA may only account for a few percent of the entire sequence). However, it is not always possible (or desirable) to only sequence the coding regions separately. Also, as scientists understand more about the role of this Non-coding DNA (often referred to as junk DNA), it will become more important to have a complete genome sequence as a background to understanding the genetics and biology of any given organism.

In many ways genome projects do not confine themselves to only determining a DNA sequence of an organism. Such projects may also include gene prediction to find out where the genes are in a genome, and what those genes do. There may also be related projects to sequence ESTs or mRNAs to help find out where the genes actually are.

Historical and technological perspectives

Historically, when sequencing eukaryotic genomes (such as the worm Caenorhabditis elegans) it was common to first map the genome to provide a series of landmarks across the genome. Rather than sequence a chromosome in one go, it would be sequenced piece by piece (with the prior knowledge of approximately where that piece is located on the larger chromosome). Changes in technology and in particular improvements to the processing power of computers, means that genomes can now be 'shotgun sequenced' in one go (there are caveats to this approach though when compared to the traditional approach).

Improvements in DNA sequencing technology has meant that the cost of sequencing a new genome sequence has steadily fallen (in terms of cost per base pair) and newer technology has also meant that genomes can be sequenced far more quickly. When research agencies decide what new genomes to sequence, the emphasis has been on species which have either a relevance to human health (*e.g.* pathogenic bacteria or vectors of disease such as mosquitos) or species which have commercial importance (*e.g.* livestock and crop plants). Secondary emphasis is placed on species whose genomes will help answer important questions in molecular evolution (*e.g.* the common chimpanzee).

In the future, it is likely that it will become even cheaper and quicker to sequence a genome. This will allow for complete genome sequences to be determined from many different individuals of the same species. For humans, this will allow us to better understand aspects of human genetic diversity.

MUTATIONS AND EVOLUTION

Every organism has a set of genes and half of the genes of that organism come from each parent. The combinations of the genes causes the variation of individuals within the species. The genes of a butterfly, an ape or a fowl carry the code that determines the appearance and the character of the butterfly, the ape and the fowl. The genetic code allows an overwhelming variety within the species of the kind. A mutation is basically a gene that has an abnormality in relation to its normal configuration. The abnormality can then be passed to successive offspring, thereby producing a marked difference. There are many different types of mutations. The smallest possible genetic mutation is a 'point-mutation'. This occurs in the DNA when the base-pairs combine with the 'wrong' partner. Multiple point-mutations are common and are found to increase substantially by the effect of mutagens. Mutations are, of course, heritable and these can extend to whole or part chromosomal mutations. Because many genes are affected by a chromosomal mutation, these often have drastic ramifications on the offspring.

Mutation

In biology, mutations are changes to the base pair sequence of genetic material (either DNA or RNA). Mutations can be caused by copying errors in the genetic material during cell division and by exposure to ultraviolet or ionizing radiation, chemical mutagens, or viruses, or can occur deliberately under cellular control during processes such as meiosis or hypermutation. In multicellular organisms, mutations can be subdivided into germline mutations, which can be passed on to descendants, and somatic mutations. The somatic mutations cannot be transmitted to descendants in animals. Plants sometimes can transmit somatic mutations to their descendants asexually or sexually (in case when flower buds develop in somatically mutated part of plant).

Mutations create variations in the gene pool, and the less favorable (or deleterious) mutations are removed from the gene pool by natural selection, while more favorable (beneficial or advantageous) ones tend to accumulate, resulting in evolutionary change. For example, a butterfly may develop offspring with a new mutation caused say by ultraviolet light from the sun. In most cases, this mutation is not good, since obviously there was no 'purpose' for such change at the molecular level. However, sometimes a mutation may change, say, the butterfly's colour, making it harder for predators to see it; this is an advantage and the chances of this butterfly surviving and producing its own offspring are a little better, and over time the number of butterflies with this mutation may form a large percentage of the species. Neutral mutations are defined as mutations whose effects do not influence the fitness of either the species or the individuals who make up the species. These can accumulate over time due to genetic drift. The overwhelming majority of mutations have no significant effect, since DNA repair is able to mend most changes before they become permanent mutations, and many organisms have mechanisms for eliminating otherwise permanently mutated somatic cells.

Classification

By effect on structure

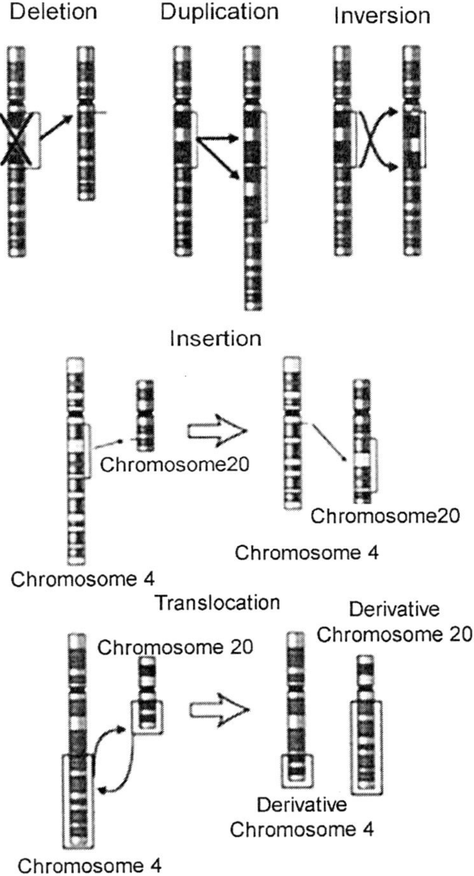

Fig. five types of chromosomal mutations.

The sequence of a gene can be altered in a number of ways. Gene mutations have varying effects on health depending on where they occur and whether they alter the function of essential proteins. Structurally, mutations can be classified as:

1. Small-scale mutations, such as affecting a small gene is one or a few nucleotides, including:

a Point mutations, often caused by chemicals or malfunction of DNA replication, exchange a single nucleotide for another. Most common is the transition that exchanges a purine for a purine (A "! G) or a pyrimidine for a pyrimidine, (C "! T). A transition can be caused by nitrous acid, base mispairing, or mutagenic base analogs such as 5-bromo-2-deoxyuridine (BrdU). Less common is a transversion, which exchanges a purine for a pyrimidine or a pyrimidine for a purine (C/T "! A/G). A point mutation can be reversed by another point mutation, in which the nucleotide is changed back to its original state (true reversion) or by second-site reversion (a complementary mutation elsewhere that results in regained gene functionality). These changes are classified as transitions or transversions. An example of a transversion is adenine (A) being converted into a cytosine (C). There are also many other examples that can be found. Point mutations that occur within the protein coding region of a gene may be classified into three kinds, depending upon what the erroneous codon codes for:

i Silent mutations: which code for the same amino acid.

ii Missense mutations: which code for a different amino acid.

iii Non-sense mutations: which code for a stop and can truncate the protein.

b. Insertions add one or more extra nucleotides into the DNA. They are usually caused by transposable elements, or errors during replication of repeating elements (*e.g.* AT repeats). Insertions in the coding region of a gene may alter splicing of the mRNA (splice site mutation), or cause a shift in the reading frame (frameshift), both of which can significantly alter the gene product. Insertions can be reverted by excision of the transposable element.

c. Deletions remove one or more nucleotides from the DNA. Like insertions, these mutations can alter the reading frame of the gene. They are irreversible.

2. Large-scale mutations in chromosomal structure, including:

a Amplifications (or gene duplications) leading to multiple copies of all chromosomal regions, increasing the dosage of the genes located within them.

b Deletions of large chromosomal regions, leading to loss of the genes within those regions.

c Mutations whose effect is to juxtapose previously separate pieces of DNA, potentially bringing together separate genes to form functionally distinct fusion genes (*e.g.* bcr-abl). These include:

i Chromosomal translocations: interchange of genetic parts from Non-homologous chromosomes.

ii Interstitial deletions: an intra-chromosomal deletion that removes a segment of DNA from a single chromosome, thereby apposing previously distant genes. For example, cells isolated from a human astrocytoma, a type of brain tumor, were found to have a chromosomal deletion removing sequences between the "fused in glioblastoma" (fig) gene and the receptor tyrosine kinase "ros", producing a fusion protein (FIG-ROS). The abnormal FIG-ROS fusion protein has constitutively active kinase activity that causes oncogenic transformation (a transformation from normal cells to cancer cells).

iii Chromosomal inversions: reversing the orientation of a chromosomal segment.

d Loss of heterozygosity: loss of one allele, either by a deletion or recombination event, in an organism that previously had two different alleles.

By effect on function

- Loss-of-function mutations are the result of gene product having less or no function. When the allele has a complete loss of function (null allele) it is often called an amorphic mutation. Phenotypes associated with such mutations are most often recessive. Exceptions are when the organism is haploid, or when the reduced dosage of a normal gene product is not enough for a normal phenotype (this is called haploinsufficiency).
- Gain-of-function mutations change the gene product such that it gains a new and abnormal function. These mutations usually have dominant phenotypes. Often called a neomorphic mutation.
- Dominant negative mutations (also called antimorphic mutations) have an altered gene product that acts antagonistically to the wild-type allele. These mutations usually result in an altered molecular function (often inactive) and are characterised by a dominant or semi-dominant phenotype. In humans, Marfan syndrome is an example of a dominant negative mutation occurring in an autosomal dominant disease. In this condition, the defective glycoprotein product of the fibrillin gene (FBN1) antagonizes the product of the normal allele.
- Lethal mutations are mutations that lead to a phenotype incapable of effective reproduction.

By aspect of phenotype affected

- Morphological mutations usually affect the outward appearance of an individual. Mutations can change the height of a plant or change it from smooth to rough seeds.
- Biochemical mutations result in lesions stopping the enzymatic pathway. Often, morphological mutants are the direct result of a mutation due to the enzymatic pathway.

Special classes

- Conditional mutation is a mutation that has wild-type (or less severe) phenotype under certain "permissive" environmental conditions and a mutant phenotype under certain "restrictive" conditions. For example, a temperature-sensitive mutation can cause cell death at high temperature (restrictive condition), but might have no deleterious consequences at a lower temperature (permissive condition).

Causes of mutation

Two classes of mutations are spontaneous mutations (molecular decay) and induced mutations caused by mutagens.

Spontaneous mutations on the molecular level include:

- Tautomerism - A base is changed by the repositioning of a hydrogen atom.
- Depurination - Loss of a purine base (A or G).
- Deamination - Changes a normal base to an atypical base; C '! U, (which can be corrected by DNA repair mechanisms), or spontaneous deamination of 5-methycytosine (irreparable), or A '! HX (hypoxanthine).
- Transition - A purine changes to another purine, or a pyrimidine to a pyrimidine.
- Transversion - A purine becomes a pyrimidine, or vice versa.

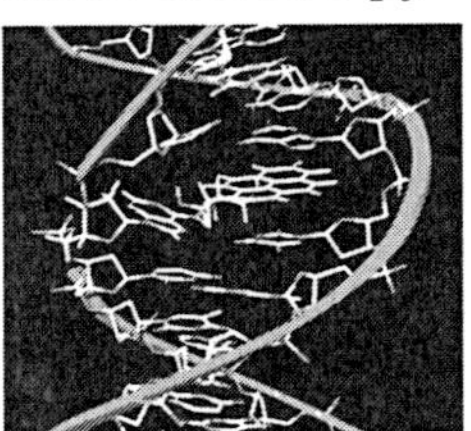

Fig. Benzopyrene, the major mutagen in tobacco smoke, in an adduct to DNA.

Induced mutations on the molecular level can be caused by:

1. Chemicals
   a. Nitrosoguanidine (NTG)
   b. Hydroxyamine NH3OH
   c. Base analogs (*e.g.* BrdU)
   d. Simple chemicals (*e.g.* acids)
   e. Alkylating agents (*e.g.* N-ethyl-N-nitrosourea (ENU)) These agents can mutate both replicating and non-replicating DNA. In contrast, a base analog can only mutate the DNA when the analog is incorporated in replicating the DNA. Each of these classes of chemical mutagens has certain effects that then lead to transitions, transversions, or deletions.
   f. Methylating agents (*e.g.* ethyl methanesulfonate (EMS))
   g. Polycyclic hydrocarbons (*e.g.* benzopyrenes found in internal

h. DNA intercalating agents (*e.g.* ethidium bromide)
i DNA crosslinker (*e.g.* platinum)
j Oxidative damage caused by oxygen(O)] radicals

1. Radiation
   a Ultraviolet radiation (nonionizing radiation) - excites electrons to a higher energy level. DNA absorbs one form, ultraviolet light. Two nucleotide bases in DNA - cytosine and thymine-are most vulnerable to excitation that can change base-pairing properties. UV light can induce adjacent thymine bases in a DNA strand to pair with each other, as a bulky dimer.
   b Ionizing radiation

DNA has so-called hotspots, where mutations occur up to 100 times more frequently than the normal mutation rate. A hotspot can be at an unusual base, *e.g.*, 5-methylcytosine.

Mutation rates also vary across species. Evolutionary biologists have theorized that higher mutation rates are beneficial in some situations, because they allow organisms to evolve and therefore adapt more quickly to their environments. For example, repeated exposure of bacteria to antibiotics, and selection of resistant mutants, can result in the selection of bacteria that have a much higher mutation rate than the original population (mutator strains).

Nomenclature

Nomenclature of mutations specify the type of mutation and base or amino acid changes.

- Amino acid substitution - (*e.g.* D111E) The first letter is the one letter code of the wildtype amino acid, the number is the position of the amino acid from the N terminus and the second letter is the one letter code of the amino acid present in the mutation. If the second letter is 'X', any amino acid may replace the wildtype.
- Amino acid deletion - (*e.g.* ÄF508) The Greek symbol Ä or 'delta' indicates a deletion. The letter refers to the amino acid present in the wildtype and the number is the position from the N terminus of the amino acid were it to be present as in the wildtype.

Harmful Mutations

Changes in DNA caused by mutation can cause errors in protein sequence, creating partially or completely non-functional proteins. To function correctly, each cell depends on thousands of proteins to function in the right places at the right times. When a mutation alters a protein that plays a critical role in the body, a medical condition can result. A condition caused by mutations in one or more genes is called a genetic disorder. However, only a small percentage of mutations cause genetic disorders; most have no impact on health. For example, some mutations alter a gene's DNA base sequence but don't change the function of the protein made by the gene.If a mutation is present in a germ cell, it can give rise to offspring that carries the mutation in all of its cells. This is the case in hereditary diseases. On the other hand, a mutation can occur in a somatic cell of an organism. Such mutations will be present in all descendants of this cell, and certain mutations can cause the cell to become malignant, and thus cause cancer.

Often, gene mutations that could cause a genetic disorder are repaired by the DNA repair system of the cell. Each cell has a number of pathways through which enzymes recognize and repair mistakes in DNA. Because DNA can be damaged or mutated in many ways, the process of DNA repair is an important way in which the body protects itself from disease.

Beneficial Mutations

A very small percentage of all mutations actually have a positive effect. These mutations lead to new versions of proteins that help an organism and its future generations better adapt to changes in their environment. For example, a specific 32 base pair deletion in human CCR5 (CCR5-32) confers HIV resistance to homozygotes and delays AIDS onset in heterozygotes. The CCR5 mutation is more common in those of European descent. One theory for the etiology of the relatively high frequency of CCR5-32 in the European population is that it conferred resistance to the bubonic plague in mid-14th century Europe.

Selection and Evolution

Mutations produce organisms with different genotypes, and those differences can result in different phenotypes. Many genetic mutations, called "neutral mutations", have a negligible effect on an organism's phenotype, health, and reproductive fitness. Mutations which do have an effect are often deleterious, but occassionally mutations arise which are beneficial in the current environmental context of the organism.

Population genetics research studies the distributions of these genetic differences within populations and how the distributions change over time. Changes in the frequency of an allele in a population can be influenced by natural selection, where a given allele's higher rate of survival and reproduction causes it to become more frequent in the population over time. Genetic drift can also occur, where chance events lead to random changes in allele frequency.

Over many generations, the genomes of organisms can change, resulting in the phenomenon of evolution. Mutations and the selection for beneficial mutations can cause a species to evolve into forms that better survive their environment, a process called adaptation. New species are formed through the process of speciation, a process often caused by geographical separations that allow different populations to genetically diverge.

As sequences diverge and change during the process of evolution, these differences between sequences can be used as a molecular clock to calculate the evolutionary distance between them. Genetic comparisons are generally considered the most accurate method of characterizing the relatedness between species, an improvement over the sometimes deceptive comparison of phenotypic characteristics.

The evolutionary distances between species can be combined to form evolutionary trees. These trees are commonly considered the most accurate representation of relatedness, although the transfer of genetic material between unrelated species (known as "horizontal gene transfer" and most common in bacteria) cannot be represented by the tree.

CHROMOSOME

The total complement of genes in an organism or cell is known as its genome, which may be stored on one or more chromosomes; the region of the chromosome at which a particular gene is located is called its locus. A chromosome consists of a single, very long DNA helix on which thousands of genes are encoded.

Prokaryotes - bacteria and archaea - typically store their genomes on a single large, circular chromosome, sometimes supplemented by additional small circles of DNA called plasmids, which usually encode only a few genes and are easily transferable between individuals. For example, the genes for antibiotic resistance are usually encoded on bacterial plasmids and can be passed between individual cells, even those of different species, via horizontal gene transfer.

Although some simple eukaryotes also possess plasmids with small numbers of genes, the majority of eukaryotic genes are stored on multiple linear chromosomes, which are packed within the nucleus in complex with storage proteins called histones. The manner in which DNA is stored on the histone, as well as chemical modifications of the histone itself, are regulatory mechanisms governing whether a particular region of DNA is accessible for gene expression. The ends of eukaryotic chromosomes are capped by long stretches of repetitive sequences called telomeres, which do not code for any gene product but are present to prevent degradation of coding and regulatory regions during DNA replication. The length of the telomeres tends to decrease each time the genome is replicated in preparation for cell division; the loss of telomeres has been proposed as an explanation for cellular senescence, or the loss of the ability to divide, and by extension for the aging process in organisms.

While the chromosomes of prokaryotes are relatively gene-dense, those of eukaryotes often contain so-called "junk DNA", or regions of DNA that serve no obvious function. Simple single-celled eukaryotes have relatively small amounts of such DNA, while the genomes of complex multicellular organisms, including humans, contain an absolute majority of DNA without an identified function.

CHROMOSOMES AND GENDER

Chromosomes are long, stringy aggregates of genes that carry heredity information. They are composed of DNA and proteins and are located within the nucleus of our cells. Chromosomes determine everything from hair colour and eye colour to gender. Whether you are a male or female depends on the presence or absence of certain chromosomes. Human cells contain 23 pairs of chromosomes for a total of 46. There are 22 pairs of autosomes and one pair of sex chromosomes. The sex chromosomes are the X chromosome and the Y chromosome. These chromosomes determine gender.

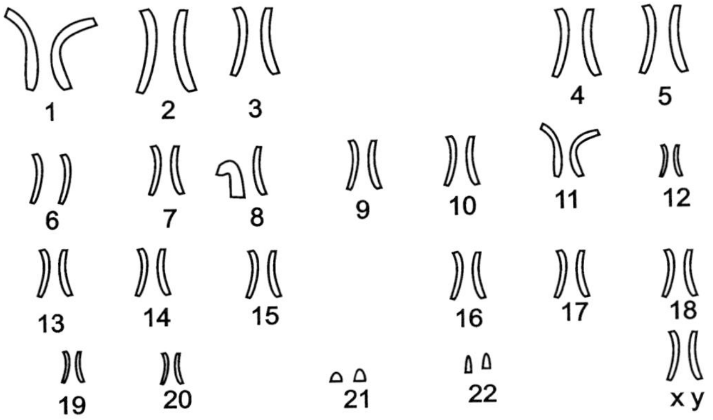

Fig. Karyotype of a normal male with 22 pairs of autosomes and one pair of sex chromosomes.

In human sexual reproduction, two distinct gametes fuse to form a zygote. Gametes are reproductive cells produced by a type of cell division called meiosis. They contain only one set of chromosomes and are said to be haploid (one set of 22 autosomes and one sex chromosome). The male gamete, called the spermatozoan, is relatively motile and usually has a flagellum. The female gamete, called the ovum, is Non-motile and relatively large in comparison to the male gamete. When the haploid male and female gametes unite in a process called fertilization, they form what is called a zygote. The zygote is diploid, meaning that it contains two sets of chromosomes (two sets of 22 autosomes and two sex chromosomes).

Sex Chromosomes X-Y

The male gametes or sperm cells in humans and other mammals are heterogametic and contain one of two types of sex chromosomes. They are either X or Y. The female gametes or eggs however, contain only the X sex chromosome and are homogametic. The sperm cell determines the sex of an individual in this case. If a sperm cell containing an X chromosome fertilizes an egg, the resulting zygote will be XX or female. If the sperm cell contains a Y chromosome, then the resulting zygote will be XY or male.

Sex Chromosomes X-O

Grasshoppers, roaches, and other insects have a similar system for determining the sex of an individual. Adult males lack a Y sex chromosome and have only an X chromosome. They produce sperm cells that contain either an X chromosome or no sex chromosome, which is designated as O. The females are XX and produce egg cells that contain an X chromosome. If an X sperm cell fertilizes an egg, the resulting zygote will be XX or female. If a sperm cell containing no sex chromosome fertilizes an egg, the resulting zygote will be XO or male.

Sex Chromosomes Z-W

Birds, insects like butterflies, and some species of fish have a different system for determining gender. In these animals it is the female gamete that determines the sex of an individual. Female gametes can either contain a Z chromosome or a W chromosome. Male gametes contain only the Z chromosome. Females of these species are ZW and males are ZZ.

Parthenogenesis

What about animals like most kinds of wasps, bees, and ants that have no sex chromosomes? How is gender determined? In these species, fertilization determines gender. If an egg becomes fertilized it will develop into a female. A non-fertilized egg may develop into a male. The female is diploid and contains two sets of chromosomes, while the male is haploid. This development of an unfertilized egg into an individual is called parthenogenesis.

11

Working of Genes

People have known for many years that living things inherit traits from their parents. That common-sense observation led to agriculture, the purposeful breeding and cultivation of animals and plants for desirable characteristics. Firming up the details took quite some time, though. Researchers did not understand exactly how traits were passed to the next generation until the middle of the 20th century.

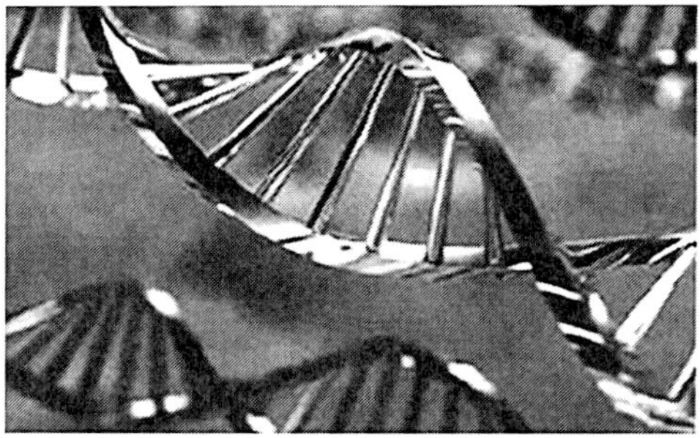

Fig. Structure of DNA.

Now it is clear that genes are what carry our traits through generations and that genes are made of deoxyribonucleic acid (DNA). But genes themselves don't do the actual work. Rather, they serve as instruction books for making functional molecules such as ribonucleic acid (RNA) and proteins, which perform the chemical reactions in our bodies.

Proteins do many other things, too. They provide the body's main building materials, forming the cell's architecture and structural components. But one thing proteins can't do is make copies of themselves. When a cell needs more proteins, it uses the manufacturing instructions coded in DNA.

The DNA code of a gene—the sequence of its individual DNA building blocks, labeled A (adenine), T (thymine), C (cytosine), and G (guanine) and collectively called nucleotides— spells out the exact order of a protein's building blocks, amino acids. Occasionally, there is a kind of typographical error in a gene's DNA sequence. This mistake— which can be a change, gap, or duplication—is called a mutation.

A mutation can cause a gene to encode a protein that works incorrectly or that doesn't work at all. Sometimes, the error means that no protein is made.

But not all DNA changes are harmful. Some mutations have no effect, and others produce new versions of proteins that may give a survival advantage to the organisms that have them. Over time, mutations supply the raw material from which new life forms evolve Beautiful DNA

Up until the 1950s, scientists knew a good deal about heredity, but they didn't have a clue what DNA looked like. In order to learn more about DNA and its structure, some scientists experimented with using X rays as a form of molecular photography.

In 1953, Watson and Crick created their historic model of the shape of DNA: the double helix.

Rosalind Franklin, a physical chemist working with Maurice Wilkins at King's College in London, was among the first to use this method to analyze genetic material. Her experiments produced what were referred to at the time as "the most beautiful X-ray photographs of any substance ever taken."

Other scientists, including zoologist James Watson and physicist Francis Crick, both working at Cambridge University in the United Kingdom, were trying to determine the shape of DNA too. Ultimately, this line of research revealed one of the most profound scientific discoveries of the 20th century: that DNA exists as a double helix.

The 1962 Nobel Prize in physiology or medicine was awarded to Watson, Crick, and Wilkins for this work. Although Franklin did not earn a share of the prize due to her untimely death at age 38, she is widely recognized as having played a significant role in the discovery.

The spiral staircase-shaped double helix has attained global status as the symbol for DNA. But what is so beautiful about the discovery of the twisting ladder structure isn't just its good looks. Rather, the structure of DNA taught researchers a fundamental lesson about genetics. It taught them that the two connected strands—winding together like parallel handrails—were complementary to each other, and this unlocked the secret of how genetic information is stored, transferred, and copied.

In genetics, complementary means that if you know the sequence of nucleotide building blocks on one strand, you know the sequence of nucleotide building blocks on the other strand: A always matches up with T and C always links to G. Long strings of nucleotides form genes, and groups of genes are packaged tightly into structures called chromosomes. Every cell in your body except for eggs, sperm, and red blood cells contains a full set of chromosomes in its nucleus.If the chromosomes in one of your cells were uncoiled and placed end to end, the DNA would be about 6 feet long. If all the DNA in your body were connected in this way, it would stretch approximately 67 billion miles! That's nearly 150,000 round trips to the Moon.

The long, stringy DNA that makes up genes is spooled within chromosomes inside the nucleus of a cell. (Note that a gene would actually be a much longer stretch of DNA than what is shown here.)

DNA consists of two long, twisted chains made up of nucleotides. Each nucleotide contains one base, one phosphate molecule, and the sugar molecule deoxyribose. The bases in DNA nucleotides are adenine, thymine, cytosine, and guanine.

Copycat

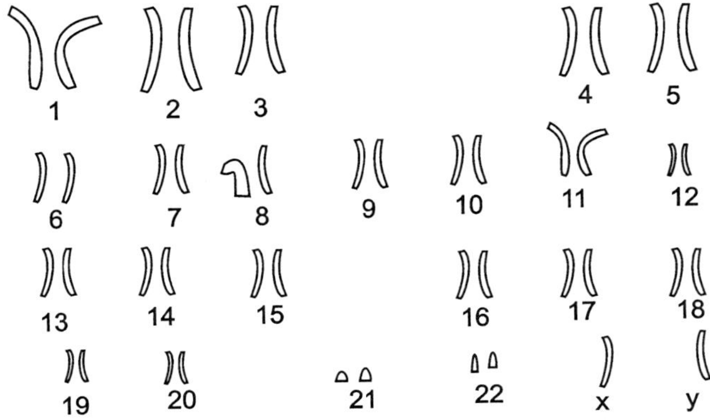

Fig. pairs of chromosomes.

Male DNA (pictured here) contains an X and a Y chromosome, whereas female DNA contains two X chromosomes.

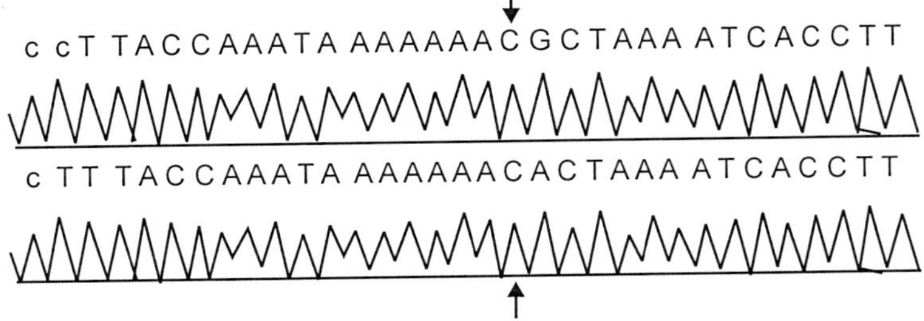

When DNA polymerase makes an error while copying a gene's DNA sequence, the mistake is called a mutation. In this example, the nucleotide G has been changed to an A. It's astounding to think that your body consists of trillions of cells. But what's most amazing is that it all starts with one cell. How does this massive expansion take place? As an embryo progresses through development, its cells must reproduce. But before a cell divides into two new, nearly identical cells, it must copy its DNA so there will be a complete set of genes to pass on to each of the new cells.

To make a copy of itself, the twisted, compacted double helix of DNA has to unwind and separate its two strands. Each strand becomes a pattern, or template, for making a new strand, so the two new DNA molecules have one new strand and one old strand. The copy is courtesy of a cellular protein machine called DNA polymerase, which reads the template DNA strand and stitches together the complementary new strand. The process, called replication, is astonishingly fast and accurate, although occasional mistakes, such as deletions or duplications, occur. Fortunately, a cellular spell-checker catches and corrects nearly all of these errors.

Mistakes that are not corrected can lead to diseases such as cancer and certain genetic disorders. Some of these include Fanconi anemia, early aging diseases, and other conditions in which people are extremely sensitive to sunlight and some chemicals.

DNA copying is not the only time when DNA damage can happen. Prolonged, unprotected sun exposure can cause DNA changes that lead to skin cancer, and toxins in cigarette smoke can cause lung cancer.

It may seem ironic, then, that many drugs used to treat cancer work by attacking DNA. That's because these chemotherapy drugs disrupt the DNA copying process, which goes on much faster in rapidly dividing cancer cells than in other cells of the body. The trouble is that most of these drugs do affect normal cells that grow and divide frequently, such as cells of the immune system and hair cells.

Understanding DNA replication better could be a key to limiting a drug's action to cancer cells only.

LET'S CALL IT EVEN

After copying its DNA, a cell's next challenge is getting just the right amount of genetic material into each of its two offspring.

Most of your cells are called diploid ("di" means two, and "ploid" refers to sets of chromosomes) because they have two sets of chromosomes (23 pairs). Eggs and sperm are different; these are known as haploid cells. Each haploid cell has only one set of 23 chromosomes so that at fertilization the math will work out: A haploid egg cell will combine with a haploid sperm cell to form a diploid cell with the right number of chromosomes: 46.

Chromosomes are numbered 1 to 22, according to size, with 1 being the largest chromosome. The 23rd pair, known as the sex chromosomes, are called X and Y. In humans, abnormalities of chromosome number usually occur during meiosis, the time when a cell reduces its chromosomes from diploid to haploid in creating eggs or sperm. What happens if an egg or a sperm cell gets the wrong number of chromosomes, and how often does this happen?

Molecular biologist Angelika Amon of the Massachusetts Institute of Technology in Cambridge says that mistakes in dividing DNA between daughter cells during meiosis are the leading cause of human birth defects and miscarriages. Current estimates are that 10 percent of all embryos have an incorrect chromosome number. Most of these don't go to full term and are miscarried.In women, the likelihood that chromosomes won't be apportioned properly increases with age. One of every 18 babies born to women over 45 has three copies of chromosome 13, 18, or 21 instead of the normal two, and this improper balancing can cause trouble. For example, three copies of chromosome 21 lead to Down syndrome.

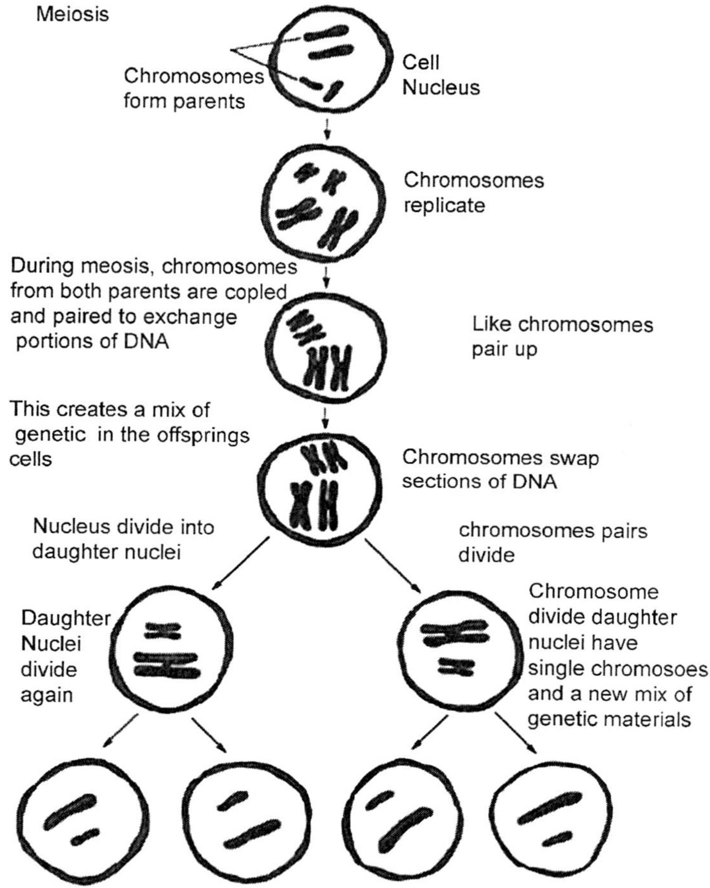

Fig .meiosis

To make her work easier, Amon—like many other basic scientists—studies yeast cells, which separate their chromosomes almost exactly the same way human cells do, except that yeast do it much faster. A yeast cell copies its DNA and produces daughter cells in about 11/2 hours, compared to a whole day for human cells.The yeast cells she uses are the same kind bakeries use to make bread and breweries use to make beer! Amon has made major progress in understanding the details of meiosis. Her research shows how, in healthy cells, gluelike protein complexes called cohesins release pairs of chromosomes at exactly the right time. This allows the chromosomes to separate properly.These findings have important implications for understanding and treating infertility, birth defects, and cancer.

GETTING THE MESSAGE

So, we've described DNA—its basic properties and how our bodies make more of it. But how does DNA serve as the language of life? How do you get a protein from a gene?

There are two major steps in making a protein. The first is transcription, where the information coded in DNA is copied into RNA. The RNA nucleotides are complementary to those on the DNA: a C on the RNA strand matches a G on the DNA strand.

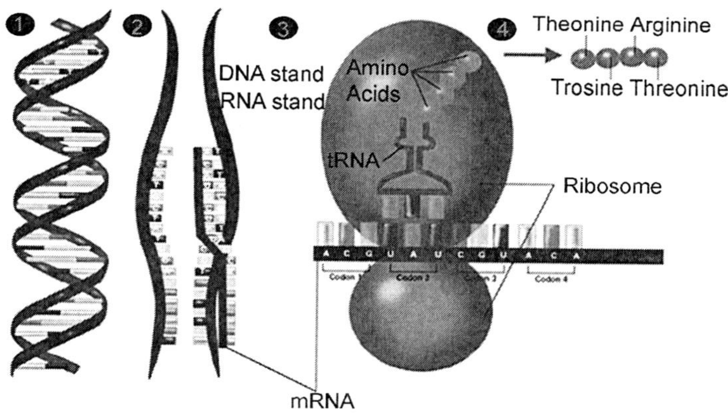

Fig. RNA polymerase transcribes DNA

RNA polymerase transcribes DNA to make messenger RNA (mRNA). The mRNA sequence (dark red strand) is complementary to the DNA sequence (blue strand). On ribosomes, transfer RNA (tRNA) helps convert mRNA into protein. Amino acids link up to make a protein.

The only difference is that RNA pairs a nucleotide called uracil (U), instead of a T, with an A on the DNA.A protein machine called RNA polymerase reads the DNA and makes the RNA copy. This copy is called messenger RNA, or mRNA, because it delivers the gene's message to the protein-producing machinery.

At this point you may be wondering why all of the cells in the human body aren't exactly alike, since they all contain the same DNA.What makes a liver cell different from a brain cell? How do the cells in the heart make the organ contract, but those in skin allow us to sweat?

That's because RNA polymerase does not work alone, but rather functions with the aid of many helper proteins. While the core part of RNA polymerase is the same in all cells, the helpers vary in different cell types throughout the body.You'd think that for a process so essential to life, researchers would know a lot about how transcription works. While it's true that the basics are clear— biologists have been studying gene transcribing by RNA polymerases since these proteins were first discovered in 1960— some of the details are actually still murky.

The biggest obstacle to learning more has been a lack of tools. Until recently, researchers were unable to get a picture at the atomic level of the giant RNA polymerase protein assemblies inside cells to understand how the many pieces of this amazing, living machine do what they do, and do it so well.

But our understanding is improving fast, thanks to spectacular technological advances. We have new X-ray pictures that are far more sophisticated than those that revealed the structure of DNA. Roger Kornberg of Stanford University in California used such methods to determine the structure of RNA polymerase. This work earned him the 2006 Nobel Prize in chemistry. In addition, very powerful microscopes and other tools that allow us to watch one molecule at a time provide a new look at RNA polymerase while it's at work reading DNA and producing RNA.

For example, Steven Block, also of Stanford, has used a physics technique called optical trapping to track RNA polymerase as it inches along DNA. Block and his team performed this work by designing a specialized microscope sensitive enough to watch the real-time motion of a single polymerase traveling down a gene on one chromosome.

The researchers discovered that molecules of RNA polymerase behave like battery-powered spiders as they crawl along the DNA ladder, adding nucleotides one at a time to the growing RNA strand. The enzyme works much like a motor, Block believes, powered by energy released during the chemical synthesis of DNA.

NATURE'S CUT-AND-PASTE JOB

Genes are often interrupted by stretches of DNA (introns, blue) that do not contain instructions for making a protein. The DNA segments that do contain protein-making instructions are known as exons (green).

Several types of RNA play key roles in making a protein. The gene transcript (the mRNA) transfers information from DNA in the nucleus to the ribosomes that make protein. Ribosomal RNA forms about 60 percent of the ribosomes. Lastly, transfer RNA carries amino acids to the ribosomes. As you can see, all three types of cellular RNAs come together to produce new proteins.

But the journey from gene to protein isn't quite as simple as we've just made it out to be. After transcription, several things need to happen to mRNA before a protein can be made. For example, the genetic material of humans and other eukaryotes (organisms that have a nucleus) includes a lot of DNA that doesn't encode proteins. Some of this DNA is stuck right in the middle of genes.

To distinguish the two types of DNA, scientists call the coding sequences of genes exons and the pieces in between introns (for intervening sequences).

If RNA polymerase were to transcribe DNA from the start of an intron-containing gene to the end, the RNA would be complementary to the introns as well as the exons.

To get an mRNA molecule that yields a working protein, the cell needs to trim out the intron sections and then stitch only the exon pieces together. This process is called RNA splicing.

Arranging exons in different patterns, called alternative splicing, enables cells to make different proteins from a single gene.

Splicing has to be extremely accurate. An error in the splicing process, even one that results in the deletion of just one nucleotide in an exon or the addition of just one nucleotide in an intron, will throw the whole sequence out of alignment. The result is usually an abnormal protein—or no protein at all. One form of Alzheimer's disease, for example, is caused by this kind of splicing error.

Molecular biologist Christine Guthrie of the University of California, San Francisco, wants to understand more fully the mechanism for removing intron RNA and find out how it stays so accurate.

She uses yeast cells for these experiments. Just like human DNA, yeast DNA has introns, but they are fewer and simpler in structure and are therefore easier to study. Guthrie can identify which genes are required for splicing by finding abnormal yeast cells that mangle splicing.

So why do introns exist, if they're just going to be chopped out? Without introns, cells wouldn't need to go through the splicing process and keep monitoring it to be sure it's working right.

As it turns out, splicing also makes it possible for cells to create more proteins.

Think about all the exons in a gene. If a cell stitches together exons 1, 2, and 4, leaving out exon 3, the mRNA will specify the production of a particular protein. But instead, if the cell stitches together exons 1, 2, and 3, this time leaving out exon 4, then the mRNA will be translated into a different protein.

By cutting and pasting the exons in different patterns, which scientists call alternative splicing, a cell can create different proteins from a single gene. Alternative splicing is one of the reasons why human cells, which have about 25,000 genes, can make hundreds of thousands of different proteins.

ALL TOGETHER NOW

Until recently, researchers looked at genes, and the proteins they encode, one at a time. Now, they can look at how large numbers of genes and proteins act, as well as how they interact. This gives them a much better picture of what goes on in a living organism.

Already, scientists can identify all of the genes that are transcribed in a cell—or in an organ, like the heart. And although researchers can't tell you, right now, what's going on in every cell of your body while you read a book or walk down the street, they can do this sort of "whole-body" scan for simpler, single-celled organisms like yeast.

Using a new technique called genome-wide location analysis, Richard Young of the Massachusetts Institute of Technology unraveled a "regulatory code" of living yeast cells, which have more than 6,000 genes in their genome. Young's technique enabled him to determine the exact places where RNA polymerase's helper proteins sit on DNA and tell RNA polymerase to begin transcribing a gene.

Since he did the experiment with the yeast exposed to a variety of different conditions, Young was able to Fig out how transcription patterns differ when the yeast cell is under stress (say, in a dry environment) or thriving in a sugary-rich nutrient solution. Done one gene at a time, using methods considered state-of-the-art just a few years ago, this kind of analysis would have taken hundreds of years.

After demonstrating that his technique worked in yeast, Young then took his research a step forward. He used a variation of the yeast method to scan the entire human genome in small samples of cells taken from the pancreases and livers of people with type 2 diabetes. He used the results to identify genes that aren't transcribed correctly in people with the disease.

This information provides researchers with an important tool for understanding how diabetes and other diseases are influenced by defective genes. By building models to predict how genes respond in diverse situations, researchers may be able to learn how to stop or jump-start genes on demand, change the course of a disease, or prevent it from ever

NURSERY GENETICS

While most genetic research uses lab organisms, test tubes, and petri dishes, the results have real consequences for people. Your first encounter with genetic analysis probably happened shortly after you were born, when a doctor or nurse took a drop of blood from the heel of your tiny foot.

Lab tests performed with that single drop of blood can diagnose certain rare genetic disorders as well as metabolic problems like phenylketonuria (PKU).

Screening newborns in this way began in the 1960s in Massachusetts with testing for PKU, a disease affecting 1 in 14,000 people. PKU is caused by an enzyme that doesn't work properly due to a genetic mutation. Those born with this disorder cannot metabolize the amino acid phenylalanine, which is present in many foods. Left untreated, PKU can lead to mental retardation and neurological damage, but a special diet can prevent these outcomes. Testing for this condition has made a huge difference in many lives. Newborn screening is governed by individual states. That means that the state in which a baby is born determines the genetic conditions for which he or she will be screened. Currently, some states test for fewer than 10 conditions, whereas others test for more than 30. All states test for PKU.

Although expanded screening for genetic diseases in newborns is advocated by some, others question the value of screening for conditions that are currently untreatable. Another issue is that some children with mild versions of certain genetic diseases may be treated needlessly.

Based on recommendations released in 2005 by the American College of Medical Genetics, government policymakers are now considering a proposal that would establish a standard, national set of newborn tests for 29 conditions, ranging from relatively common hearing problems to very rare metabolic diseases.

TRANSLATION

After a gene has been read by RNA polymerase and the RNA is spliced, what happens next in the journey from gene to protein? The next step is reading the RNA information and fitting the building blocks of a protein together. This is called translation, and its principal actors are the ribosome and amino acids.

Ribosomes are among the biggest and most intricate structures in the cell. The ribosomes of bacteria contain not only huge amounts of RNA, but also more than 50 different proteins. Human ribosomes have even more RNA and between 70 and 80 different proteins!

Harry Noller of the University of California, Santa Cruz, has found that a ribosome performs several key jobs when it translates the genetic code of mRNA. As the messenger RNA threads through the ribosome protein machine, the ribosome reads the mRNA sequence and helps recognize and recruit the correct amino acid-carrying transfer RNA to match the mRNA code. The ribosome also links each additional amino acid into a growing protein chain.

For many years, researchers believed that even though RNAs formed a part of the ribosome, the protein portion of the ribosome did all of the work. Noller thought, instead, that maybe RNA, not proteins, performed the ribosome's job. His idea was not popular at first, because at that time it was thought that RNA could not perform such complex functions.

Some time later, however, the consensus changed. Sidney Altman of Yale University in New Haven, Connecticut, and Thomas Cech, who was then at the University of Colorado in Boulder, each discovered that RNA can perform work as complex as that done by protein enzymes. Their "RNA-as-an-enzyme" discovery turned the research world on its head and earned Cech and Altman the 1989 Nobel Prize in chemistry. Noller and other researchers have continued the painstaking work of understanding ribosomes. In 1999, he showed how different parts of a bacterial ribosome interact with one another and how the ribosome interacts with molecules involved in protein synthesis. These studies provided near proof that the fundamental mechanism of translation is performed by RNA, not by the proteins of the ribosome.

RNA SURPRISES

But which ribosomal RNAs are doing the work? Most scientists assumed that RNA nucleotides buried deep within the ribosome complex—the ones that have the same sequence in every species from bacteria to people—were the important ones for piecing the growing protein together.

However, recent research by Rachel Green, who worked with Noller before moving to Johns Hopkins University in Baltimore, Maryland, showed that this is not the case. Green discovered that those RNA nucleotides are not needed for assembling a protein. Instead, she found, the nucleotides do something else entirely: They help the growing protein slip off the ribosome once it's finished.

Noller, Green, and hundreds of other scientists work with the ribosomes of bacteria. Why should you care about how bacteria create proteins from their genes?

One reason is that this knowledge is important for learning how to disrupt the actions of disease-causing microorganisms. For example, antibiotics like erythromycin and neomycin work by attacking the ribosomes of bacteria, which are different enough from human ribosomes that our cells are not affected by these drugs.As researchers gain new information about bacterial translation, the knowledge may lead to more antibiotics for people.

New antibiotics are urgently needed because many bacteria have developed resistance to the current arsenal. This resistance is sometimes the result of changes in the bacteria's ribosomal RNA. It can be difficult to find those small, but critical, changes that may lead to resistance, so it is important to find completely new ways to block bacterial translation.

Green is working on that problem too. Her strategy is to make random mutations to the genes in a bacterium that affect its ribosomes. But what if the mutation disables the ribosome so much that it can't make proteins? Then the bacterium won't grow, and Green wouldn't find it.

Using clever molecular tricks, Green Figd out a way to rescue some of the bacteria with defective ribosomes so they could grow. While some of the rescued bacteria have changes in their ribosomal RNA that make them resistant to certain antibiotics (and thus would not make good antibiotic targets) other RNA changes that don't affect resistance may point to promising ideas for new antibiotics.

AN INTERESTING DEVELOPMENT

In the human body, one of the most important jobs for proteins is to control how embryos develop. Scientists discovered a hugely important set of proteins involved in development by studying mutations that cause bizarre malformations in fruit flies.

The most famous such abnormality is a fruit fly with a leg, rather than the usual antenna, growing out of its head. According to Thomas C. Kaufman of Indiana University in Bloomington, the leg is perfectly normal—it's just growing in the wrong place.

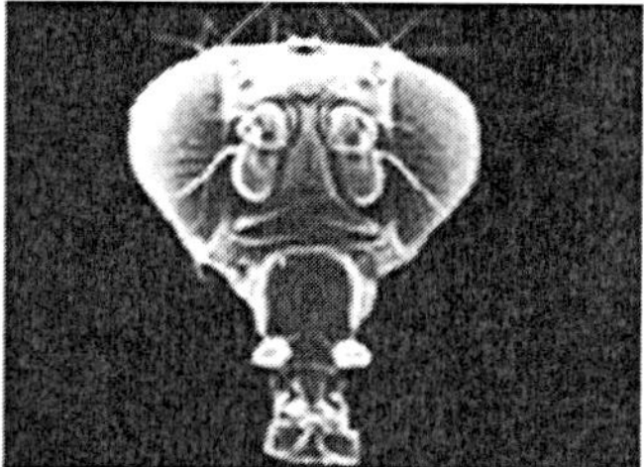

Fig. Normal fruit fly head.

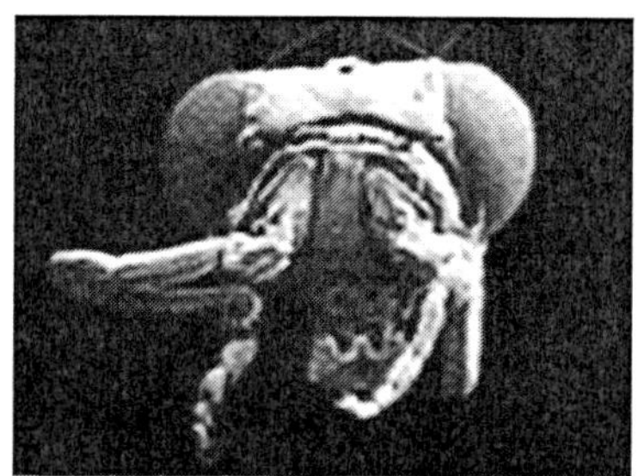

Fig. Fruit fly head showing the effects of the Antennapedia gene. This fly has legs where its antennae should be.

In this type of mutation and many others, something goes wrong with the genetic programme that directs some of the cells in an embryo to follow developmental pathways, which are a series of chemical reactions that occur in a specific order. In the antenna-into-leg problem, it is as if the cells growing from the fly's head, which normally would become an antenna, mistakenly believe that they are in the fly's thorax, and therefore ought to grow into a leg. And so they do.

Thinking about this odd situation taught scientists an important lesson—that the proteins made by some genes can act as switches. Switch genes are master controllers that provide each body part with a kind of identification card. If a protein that normally instructs cells to become an antenna is disrupted, cells can receive new instructions to become a leg instead.

Scientists determined that several different genes, each with a common sequence, provide these anatomical identification card instructions. Kaufman isolated and described one of these genes, which became known as Antennapedia, a word that means "antenna feet."

Kaufman then began looking a lot more closely at the molecular structure of the Antennapedia gene. In the early 1980s, he and other researchers made a discovery that has been fundamental to understanding evolution as well as developmental biology. The scientists found a short sequence of DNA, now called the homeobox, that is present not only in Antennapedia but in the several genes next to it and in genes in many other organisms.When geneticists find very similar DNA sequences in the genes of different organisms, it's a good clue that these genes do something so important and useful that evolution uses the same sequence over and over and permits very few changes in its structure as new species evolve.

Researchers quickly discovered nearly identical versions of homeobox DNA in almost every non-bacterial cell they examined—from yeast to plants, frogs, worms, beetles, chickens, mice, and people.

Hundreds of homeobox-containing genes have been identified, and the proteins they make turn out to be involved in the early stages of development of many species. For example, researchers have found that abnormalities in the homeobox genes can lead to extra fingers or toes in humans.

RNA AND DNA REVEALED: NEW ROLES, NEW RULES

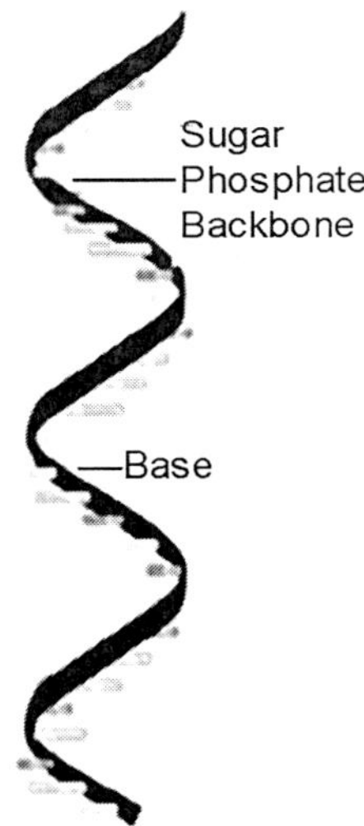

Fig. Ribonucleic acid (RNA) has the bases adenine (A), cytosine (C), guanine (G), and uracil (U).

For many years, when scientists thought about heredity, DNA was the first thing to come to mind. It's true that DNA is the basic ingredient of our genes and, as such, it often steals the limelight from RNA, the other form of genetic material inside our cells.But, while they are both types of genetic material, RNA and DNA are rather different. The chemical units of RNA are like those of DNA, except that RNA has the nucleotide uracil (U) instead of thymine (T). Unlike double-stranded DNA, RNA usually comes as only a single strand. And the nucleotides in RNA contain ribose sugar molecules in place of deoxyribose. RNA is quite flexible—unlike DNA, which is a rigid, spiral-staircase molecule that is very stable. RNA can twist itself into a variety of complicated, three-dimensional shapes. RNA is also unstable in that cells constantly break it down and must continually make it fresh, while DNA is not broken down often. RNA's instability lets cells change their patterns of protein synthesis very quickly in response to what's going on around them.

Many textbooks still portray RNA as a passive molecule, simply a "middle step" in the cell's gene-reading activities. But that view is no longer accurate. Each year, researchers unlock new secrets about RNA. These discoveries reveal that it is truly a remarkable molecule and a multi-talented actor in heredity.

Today, many scientists believe that RNA evolved on the Earth long before DNA did. Researchers hypothesize—obviously, no one was around to write this down—that RNA was a major participant in the chemical reactions that ultimately spawned the first signs of life on the planet.

RNA WORLD

At least two basic requirements exist for making a cell: the ability to hook molecules together and break them apart, and the ability to replicate, or copy itself, from existing information.

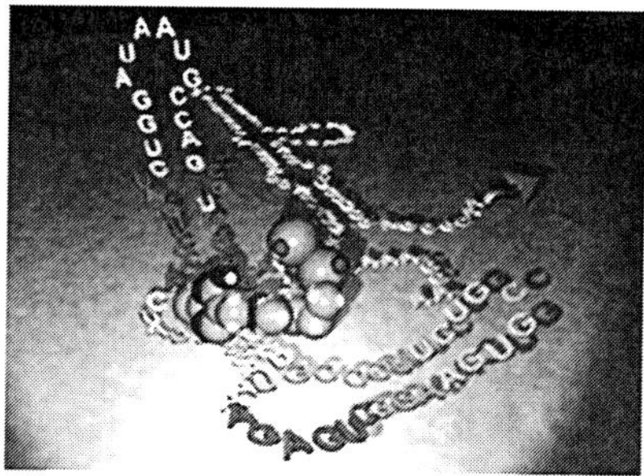

Fig. Riboswitches

Riboswitches are RNA sequences that control gene activity. The riboswitch shown here bends into a special shape when it grips tightly onto a molecule called a metabolite (colored balls) that bacteria need to survive.

RNA probably helped to form the first cell. The first organic molecules, meaning molecules containing carbon, most likely arose out of random collisions of gases in the Earth's primitive atmosphere, energy from the Sun, and heat from naturally occurring radioactivity. Some scientists think that in this primitive world, RNA was a critical molecule because of its ability to lead a double life: to store information and to conduct chemical reactions. In other words, in this world, RNA served the functions of both DNA and proteins.What does any of this have to do with human health? Plenty, it turns out.Today's researchers are harnessing some of RNA's flexibility and power. For example, through a strategy he calls directed evolution, molecular engineer Ronald R. Breaker of Yale University is developing ways to create entirely new forms of RNA and DNA that both work as enzymes.

Recently, Breaker and others have also uncovered a hidden world of RNAs that play a major role in controlling gene activity, a job once thought to be performed exclusively by proteins. These RNAs, which the scientists named riboswitches, are found in a wide variety of bacteria and other organisms.

This discovery has led Breaker to speculate that new kinds of antibiotic medicines could be developed to target bacterial riboswitches.

Molecular Editor

Fig.RNA comes in a variety of different shapes. Double-stranded DNA is a staircase-like molecule.

Scientists are learning of another way to customise proteins: by RNA editing. Although DNA sequences spell out instructions for producing RNA and proteins, these instructions aren't always followed precisely. Editing a gene's mRNA, even by a single chemical letter, can radically change the resulting protein's function. Nature likely evolved the RNA editing function as a way to get more proteins out of the same number of genes. For example, researchers have found that the mRNAs for certain proteins important for the proper functioning of the nervous system are particularly prone to editing. It may be that RNA editing gives certain brain cells the capacity to react quickly to a changing environment.

Which molecules serve as the editor and how does this happen? Brenda Bass of the University of Utah School of Medicine in Salt Lake City studies one particular class of editors called adenosine deaminases. These enzymes "retype" RNA letters at various places within an mRNA transcript.They do their job by searching for characteristic RNA shapes. Telltale twists and bends in folded RNA molecules signal these enzymes to change the RNA sequence, which in turn changes the protein that gets made.Bass' experiments show that RNA editing occurs in a variety of organisms, including people. Another interesting aspect of editing is that certain disease-causing microorganisms, such as some forms of parasites, use RNA editing to gain a survival edge when living in a human host. Understanding the details of this process is an important area of medical research.

MICRO RNA

Recently, molecules called microRNAs have been found in organisms as diverse as plants, worms, and people. The molecules are truly "micro," consisting of only a few dozen nucleotides, compared to typical human mRNAs that are a few thousand nucleotides long.

What's particularly interesting about microRNAs is that many of them arise from DNA that used to be considered merely filler material.

How do these small but important RNA molecules do their work? They start out much bigger but get trimmed by cellular enzymes, including one aptly named Dicer. Like tiny pieces of Velcro®, microRNAs stick to certain mRNA molecules and stop them from passing on their protein-making instructions.First discovered in a roundworm model system some microRNAs help determine the organism's body plan.

In their absence, very bad things can happen. For example, worms engineered to lack a microRNA called let-7 develop so abnormally that they often rupture and practically break in half as the worm grows.Perhaps it is not surprising that since microRNAs help specify the timing of an organism's developmental plan, the appearance of the microRNAs themselves is carefully timed inside a developing organism. Biologists, including Amy Pasquinelli of the University of California, San Diego, are currently figuring out how microRNAs are made and cut to size, as well as how they are produced at the proper time during development.

MicroRNA molecules also have been linked to cancer. For example, Gregory Hannon of the Cold Spring Harbour Laboratory on Long Island, New York, found that certain microRNAs are associated with the severity of the blood cancer B-cell lymphoma in mice.Since the discovery of microRNAs in the first years of the 21st century, scientists have identified hundreds of them that likely exist as part of a large family with similar nucleotide sequences. New roles for these molecules are still being found.

Healthy Interference

RNA controls genes in a way that was only discovered recently: a process called RNA interference, or RNAi. Although scientists identified RNAi less than 10 years ago, they now know that organisms have been using this trick for millions of years.

Fig. Gene Encoded

Researchers believe that RNAi arose as a way to reduce the production of a gene's encoded protein for purposes of fine-tuning growth or self-defence. When viruses infect cells, for example, they command their host to produce specialized RNAs that allow the virus to survive and make copies of itself. Researchers believe that RNAi eliminates unwanted viral RNA, and some speculate that it may even play a role in human immunity.

Oddly enough, scientists discovered RNAi from a failed experiment! Researchers investigating genes involved in plant growth noticed something strange: When they tried to turn petunia flowers purple by adding an extra "purple" gene, the flowers bloomed white instead.This result fascinated researchers, who could not understand how adding genetic material could somehow get rid of an inherited trait. The mystery remained unsolved until, a few years later, two geneticists studying development saw a similar thing happening in lab animals.The researchers, Andrew Z. Fire, then of the Carnegie Institution of Washington in Baltimore and now at Stanford University, and Craig Mello of the University of Massachusetts Medical School in Worcester, were trying to block the expression of genes that affect cell growth and tissue formation in roundworms, using a molecular tool called antisense RNA.

To their surprise, Mello and Fire found that their antisense RNA tool wasn't doing much at all. Rather, they determined, a doublestranded contaminant produced during the synthesis of the single-stranded antisense RNA interfered with gene expression. Mello and Fire named the process RNAi, and in 2006 were awarded the Nobel Prize in physiology or medicine for their discovery.Further experiments revealed that the doublestranded RNA gets chopped up inside the cell into much smaller pieces that stick to mRNA and block its action, much like the microRNA pieces of Velcro discussed above.Today, scientists are taking a cue from nature and using RNAi to explore biology. They have learned, for example, that the process is not limited to worms and plants, but operates in humans too.Medical researchers are currently testing new types of RNAi-based drugs for treating conditions such as macular degeneration, the leading cause of blindness, and various infections, including those caused by HIV and herpes virus.

Dynamic DNA

A good part of who we are is "written in our genes," inherited from Mom and Dad. Many traits, like red or brown hair, body shape, and even some personality quirks, are passed on from parent to offspring.But genes are not the whole story. Where we live, how much we exercise, what we eat: These and many other environmental factors can all affect how our genes get expressed.You know that changes in DNA and RNA can produce changes in proteins. But additional control happens at the level of DNA, even though these changes do not alter DNA directly. Inherited factors that do not change the DNA sequence of nucleotides are called epigenetic changes, and they too help make each of us unique.

Epigenetic means, literally, “upon” or “over” genetics. It describes a type of chemical reaction that can alter the physical properties of DNA without changing its sequence. These changes make genes either more or less likely to be expressed.Currently, scientists are following an intriguing course of discovery to identify epigenetic factors that, along with diet and other environmental influences, affect who we are and what type of illnesses we might get.

Secret Code

DNA is spooled up compactly inside cells in an arrangement called chromatin. This packaging is critical for DNA to do its work. Chromatin consists of long strings of DNA spooled around a compact assembly of proteins called histones.

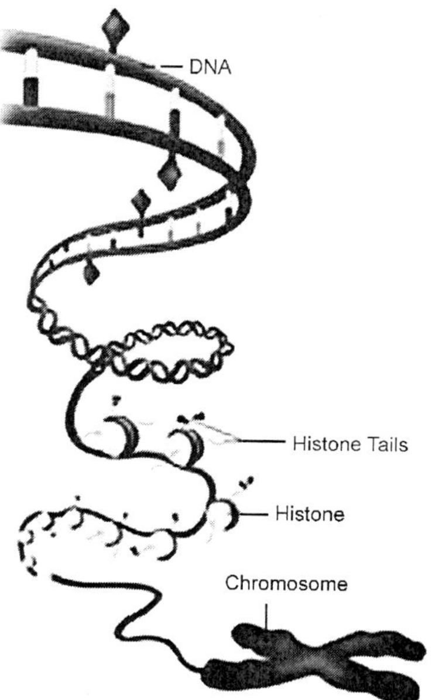

Fig. The epigenetic code

The "epigenetic code" controls gene activity with chemical tags that mark DNA (purple diamonds) and the "tails" of histone proteins (purple triangles). These markings help determine whether genes will be transcribed by RNA polymerase. Genes hidden from access to RNA polymerase are not expressed.

One of the key functions of chromatin is to control access to genes, since not all genes are turned on at the same time. Improper expression of growth-promoting genes, for example, can lead to cancer, birth defects, or other health concerns. Many years after the structure of DNA was determined, researchers used a powerful device known as an electron microscope to take pictures of chromatin fibers. Upon viewing chromatin up close, the researchers described it as "beads on a string," an image still used today. The beads were the histone balls, and the string was DNA wrapped around the histones and connecting one bead to the next.Decades of study eventually revealed that histones have special chemical tags that act like switches to control access to the DNA. Flipping these switches, called epigenetic markings, unwinds the spooled DNA so the genes can be transcribed.The observation that a cell's gene-reading machinery tracks epigenetic markings led C. David Allis, who was then at the University of Virginia Health Sciences Centre in Charlottesville and now works at the Rockefeller University in New York City, to coin a new phrase, the "histone code." He and others believe that the histone code plays a major role in determining which proteins get made in a cell.Flaws in the histone code have been associated with several types of cancer, and researchers are actively pursuing the development of medicines to correct such errors.

BATTLE OF THE SEXES

A process called imprinting, which occurs naturally in our cells, provides another example of how epigenetics affects gene activity.

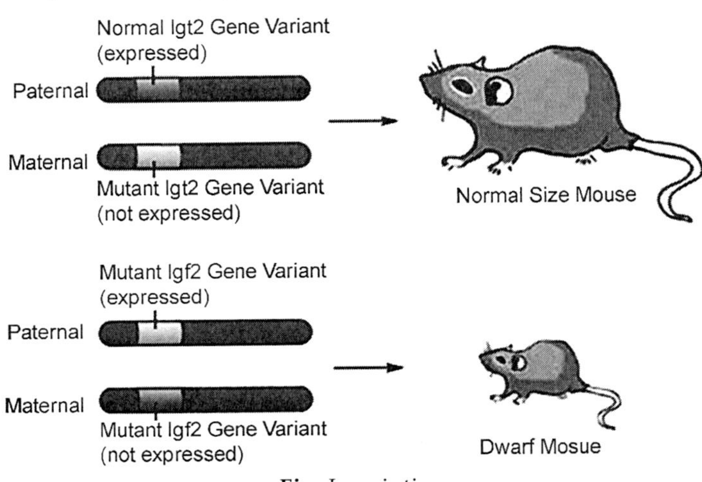

Fig. Imprinting

Igf2 is an imprinted gene. A single copy of the abnormal, or mutant, form of the Igf2 gene (red) causes growth defects, but only if the abnormal gene variant is inherited from the father.

With most genes, the two copies work exactly the same way. For some mammalian genes, however, only the mother's or the father's copy is switched on regardless of the child's gender. This is because the genes are chemically marked, or imprinted, during the process that generates eggs and sperm.As a result, the embryo that emerges from the joining of egg and sperm can tell whether a gene copy came from Mom or Dad, so it knows which copy of the gene to shut off.

One example of an imprinted gene is insulin-like growth factor 2 (Igf2), a gene that helps a mammalian fetus grow. In this case, only the father's copy of Igf2 is expressed, and the mother's copy remains silent (is not expressed) throughout the life of the offspring.

Scientists have discovered that this selective silencing of Igf2 and many other imprinted genes occurs in all placental mammals (all except the platypus, echidna, and marsupials) examined so far, but not in birds.

Why would nature tolerate a process that puts an organism at risk because only one of two copies of a gene is working? The likely reason, many researchers believe, is that mothers and fathers have competing interests, and the battle-field is DNA!The scenario goes like this: It is in a father's interest for his embryos to get bigger faster, because that will improve his offspring's chances of survival after birth. The better an individual's chance of surviving infancy, the better its chance of becoming an adult, mating, and passing its genes on to the next generation.Of course mothers want strong babies, but unlike fathers, mothers provide physical resources to embryos during pregnancy. Over her lifetime, a female is likely to be pregnant several times, so she needs to divide her resources among a number of embryos in different pregnancies.

Researchers have discovered over 200 imprinted genes in mammals since the first one was identified in 1991. We now know that imprinting controls some of the genes that have an important role in regulating embryonic and fetal growth and allocating maternal resources. Not surprisingly, mutations in these genes cause serious growth disorders.Marisa Bartolomei of the University of Pennsylvania School of Medicine in Philadelphia is trying to Fig out how Igf2 and other genes become imprinted and stay silent throughout the life of an individual. She has already identified sequences within genes that are essential for imprinting. Bartolomei and other researchers have shown that these sequences, called insulators, serve as "landing sites" for a protein that keeps the imprinted gene from being transcribed.

Starting at the End

When we think of DNA, we think of genes. However, some DNA sequences are different: They don't encode RNAs or proteins. Introns, described earlier are in this category.

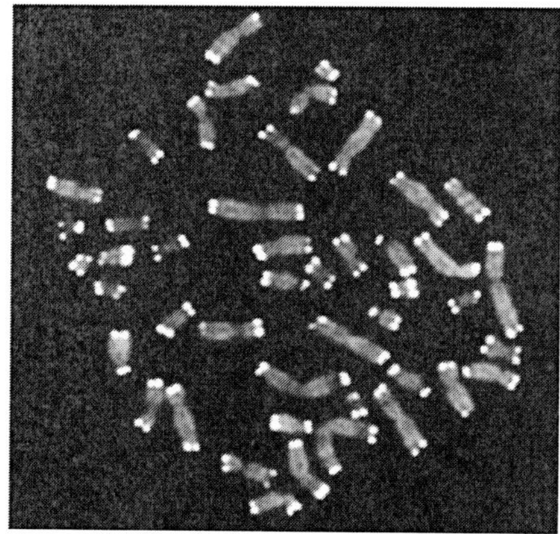

Fig. Telomeres, repeated nucleotide sequences at the tips of chromosomes, appear white in this figure.

Another example is telomeres—the ends of chromosomes. There are no genes in telomeres, but they serve an essential function. Like shoelaces without their tips, chromosomes without telomeres unravel and fray. And without telomeres, chromosomes stick to each other and cause cells to undergo harmful changes like dividing abnormally.

Researchers know a good deal about telomeres, dating back to experiments performed in the 1970s by Elizabeth Blackburn, a basic researcher who was curious about some of the fundamental events that take place within cells.

At the time, Blackburn, now at the University of California, San Francisco, was working with Joseph Gall at Yale University. For her experimental system, she chose a single-celled, pond-dwelling organism named Tetrahymena. These tiny, pear-shaped creatures are covered with hairlike cilia that they use to propel themselves through the water as they devour bacteria and fungi.

Tetrahymena was a good organism for Blackburn's experiments because it has a large number of chromosomes—which means it has a lot of telomeres!

Her research was also perfectly timed, because methods for sequencing DNA were just being developed. Blackburn found that Tetrahymena's telomeres had an unusual nucleotide sequence: TTGGGG, repeated about 50 times per telomere.

Since then, scientists have discovered that the telomeres of almost all organisms have repeated sequences of DNA with lots of Ts and Gs. In human and mouse telomeres, for example, the repeated sequence is TTAGGG.

The number of telomere repeats varies enormously, not just from organism to organism but in different cells of the same organism and even within a single cell over time. Blackburn reasoned that the repeat number might vary if cells had an enzyme that added copies of the repeated sequence to the telomeres of some but not all chromosomes.With her then-graduate student Carol Greider, now at Johns Hopkins University, Blackburn hunted for the enzyme. The team found it and Greider named it telomerase.

The telomerase enzyme, it turns out, consisted of a protein and an RNA component, which the enzyme uses as a template for copying the repeated DNA sequence.

What is the natural function of telomerase? As cells divide again and again, their telomeres get shorter.Most normal cells stop dividing when their telomeres wear down to a certain point, and eventually the cells die. Telomerase can counteract the shortening. By adding DNA to telomeres, telomerase rebuilds the telomere and resets the cell's molecular clock.

The discovery of telomerase triggered new ideas and literally thousands of new studies. Many researchers thought that the enzyme might play important roles in cancer and aging. Researchers were hoping to find ways to turn telomerase on so that cells would continue to divide (to grow extra cells for burn patients, for example), or off so that cells would stop dividing (to stop cancer, for instance).

So far, they have been unsuccessful. Although it is clear that telomerase and cellular aging are related, researchers do not know whether telomerase plays a role in the normal cellular aging process or in diseases like cancer.

Recently, however, Blackburn and a team of other scientists discovered that chronic stress and the perception that life is stressful affect telomere length and telomerase activity in the cells of healthy women. Blackburn and her coworkers are currently conducting a long-term, follow-up study to confirm these intriguing results.

THE OTHER HUMAN GENOME

Before you think everything's been said about DNA, there's one little thing we didn't mention: Some of the DNA in every cell is quite different from the DNA that we've been talking about up to this point. This special DNA isn't in chromosomes— it isn't even inside the cell's nucleus where all the chromosomes are!

So where is this special DNA? It's inside mitochondria, the organelles in our cells that produce the energy-rich molecule adenosine triphosphate, or ATP. Mendel knew nothing of mitochondria, since they weren't discovered until late in the 19th century. And it wasn't until the 1960s that researchers discovered the mitochondrial genome, which is circular like the genomes of bacteria.

In human cells, mitochondrial DNA makes up less than 1 percent of the total DNA in each of our cells. The mitochondrial genome is very small—containing only about three dozen genes. These encode a few of the proteins that are in the mitochondrion, plus a set of ribosomal RNAs used for synthesizing proteins for the organelle.Mitochondria need many more proteins though, and most of these are encoded by genes in the nucleus. Thus, the energy-producing capabilities of human mitochondria—a vital part of any cell's everyday health—depend on coordinated teamwork among hundreds of genes in two cellular neighborhoods: the nucleus and the mitochondrion.

Mitochondrial DNA gets transcribed and the RNA is translated by enzymes that are very different from those that perform this job for genes in our chromosomes. Mitochondrial enzymes look and act much more like those from bacteria, which is not surprising because mitochondria are thought to have descended from free-living bacteria that were engulfed by another cell over a billion years ago. Scientists have linked mitochondrial DNA defects with a wide range of age-related diseases including neurodegenerative disorders, some forms of heart disease, diabetes, and various cancers. It is still unclear, though, whether damaged mitochondria are a symptom or a cause of these health conditions.

Scientists have studied mitochondrial DNA for another reason: to understand the history of the human race. Unlike our chromosomal DNA, which we inherit from both parents, we get all of our mitochondrial DNA from our mothers.

Thus, it is possible to deduce who our maternal ancestors were by tracking the inheritance of mutations in mitochondrial DNA. For reasons that are still not well understood, mutations accumulate in mitochondrial DNA very slowly compared to chromosomal DNA. So, it's possible to trace your maternal ancestry way back beyond any relatives you may know by name—all the way back to "African Eve," the ancestor of us all!

Bibliography

A J Grove and G E Newell : *Animal Biology*, Shubhi Publication, Delhi, 2007.

Amita Sarkar : *A Textbook of Human Genetics*, Wisdom Press, Delhi, 2011.

Andrew L. Winton and Kate Barber Winton : *Milk Butter Cheese Ice Cream Eggs Meat Meat Extracts Gelatin Animal Fats Poultry Fish*, Agrobios Publication, Delhi, 2013.

Ashok Ganguli : *Applied Biotechnology and Plant Genetics*, Oxford Book Company, Delhi, 2009.

B.D. Pandey and B.D. Joshi : *Bioresources for Rural Livelihood, Vol. I. Genetics*, Narendra Publication, Delhi, 2010.

B.D. Sharma and J.C. Rana.: *Plant Genetic Resources of Western Himalayas: Status and Prospects*, Bishen Singh Mahendra Pal Singh, Dehradun, 2005.

B.P. Singh and Umesh Srivastava.: *Plant Genetic Resources in Indian Perspective Theory and Practices*, Indian Council of Agricultural Research, 2004.

C D N Singh : *Advanced Pathology and Treatment of Diseases of Domestic Animals*, International Book Distributing Co., 2008.

C S Patil; Prabhu M Biradar and Ajit K Gangawane: *Cell Biology*, APH Publications, Delhi, 2008.

C. LLoyd Organ : *Animal Behaviour*, MJP Publishers, Delhi, 2012.

C.H. Waddington : *An Introduction to Modern Genetics*, Vandana Publication, Delhi, 2006.

C.P. Trivedi: *Vedic Cell Biology with Life Energy and Rebirth*, Parimal Publications, Delhi, 2007.

D. Gopalakrishna Rao : *A Text Book on Tumors of Domestic Animals*, International Book, Delhi, 2004.

E A Jayson and P S Easa : *A Field Guide to Animal Signs* , Kerala Forest Research Institute, Delhi, 2004.

J.B. Jain and Sumit Jain.: *Biotech's Dictionary of Plant Breeding and Genetics*, Biotech Books, Delhi, 2005.

James A.S. Watson, Wattie J. West and C.V. Dadd.: *Crops: Varieties and Plant Breeding*, Satish Serial Publications, Delhi, 2005.

K.V. Mohanan.: *Essentials of Plant Breeding*, PHI Learning, Delhi, 2007.

Mousumi Debnath: *Molecular Cell Biology (2 Vols-Set)*, Pointer Publications, Jaipur, 2008.

Neal C. Stoskopf, Dwight T. Tomes and B.R. Christie.: *Plant Breeding: Theory and Practice*, Scientific Publications, Delhi, 2006.

O.G. Smith, K.N : *A History of Animals, Vol. I and II*, Book House, Delhi, 2009.

P C Trivedi and Indu Rani Sharma.: *Cell Biology, Genetics and Plant Breeding*, Indus Valley Publications, Jaipur, 2006.

P Ramadass : *Animal Biotechnology : Recent Concepts and Developments*, MJP Publication, Delhi, 2008.

P. Kanakaraj : *A Text Book of Animal Genetics*, International Book Distributing Company, Delhi, 2007.

P. Parvatha Reddy: *Biofumigation and Solarization for Management of Soil Borne Plant Pathogens*, Scientific Publication, Delhi, 2011.

P. Vidhyasekaran: *Bacterial Disease Resistance in Plants : Molecular Biology and Biotechnological Applications*, International Books, Delhi, 2005.

P.K.G. Nair and K. Prabhakar Achar : *A Textbook of Genetics and Evolution*, Konark Publication, 1998.

Prabhakar Sharma, Samresh Choudhuri and Abhijit Paintal: *Handbook of Cell Biology*, Wisdom Press, Delhi, 2011.

R.L. Kapoor and M.L. Saini.: *Plant Breeding and Crop Improvement (2 Vol-Set)*, CBS publications, Delhi, 1997.

S Kannaiyan and A Gopalam: *Agrobiodiversity, Vol. III. Diversity in Pests and Pathogens*, Associated Publishing Company, Delhi, 2008.

S Sundrara Rajan: *Cell and Molecular Biology*, Anmol Publication, Delhi, 2003.

Vaibhav Suri : *A Modern Book on Biological Science—Cell/Genetics*, Cyber Tech Publication, Delhi, 2009.

William J.A. Payne and R. Trevor Wilson : *An Introduction to Animal Husbandry in the Tropics*, Wiley Publication, Delhi, 2013.

Yougesh Kumar and Rajeev Tyagi : *Aquaculture Fisheries Biotechnology and Genetics*, Manglam Publishers, Delhi, 2013.